W9-ACN-477

*What bird is it that comes*
*Leaving footprints*
*On this paper strand?*

# WASHI

# THE WORLD OF JAPANESE PAPER

Sukey Hughes

KODANSHA INTERNATIONAL
Tokyo, New York and San Francisco

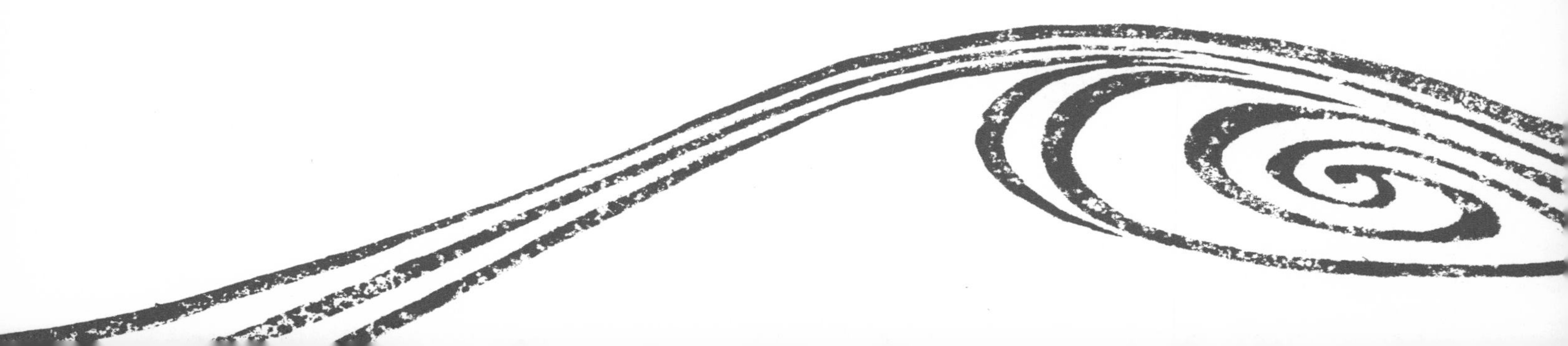

Distributed in the United States by Kodansha International/USA Ltd., through Harper & Row, Publishers, Ltd., 10 East 53rd Street, New York, New York 10022.

Published by Kodansha International Ltd., 2-12-21 Otowa, Bunkyo-ku, Tokyo 112 and Kodansha International/USA Ltd. 10 East 53rd Street, New York, New York 10022 and 44 Montgomery Street, San Francisco, California 94104. 

*LCC 77-74831*
*ISBN 0-87011-318-6*
*JBC 1072-786681-2361*

*First edition, 1978*

1. Sanuki kite, Kagawa Pref.

2. Kantō Festival, Akita Pref.

3. Nebuta Festival, Aomori Pref.

4, 5. Making umbrellas.

6. Drying tie-dyed *momigami*, Kurodani, Kyoto Pref.

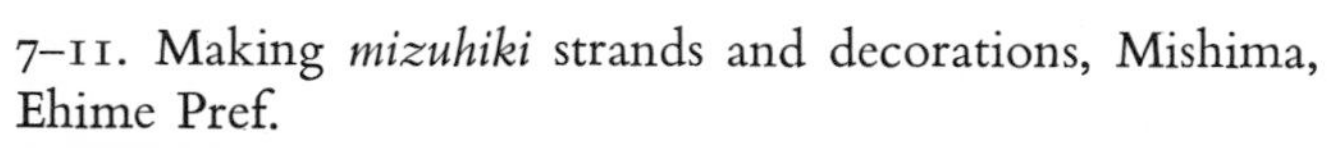

7–11. Making *mizuhiki* strands and decorations, Mishima, Ehime Pref.

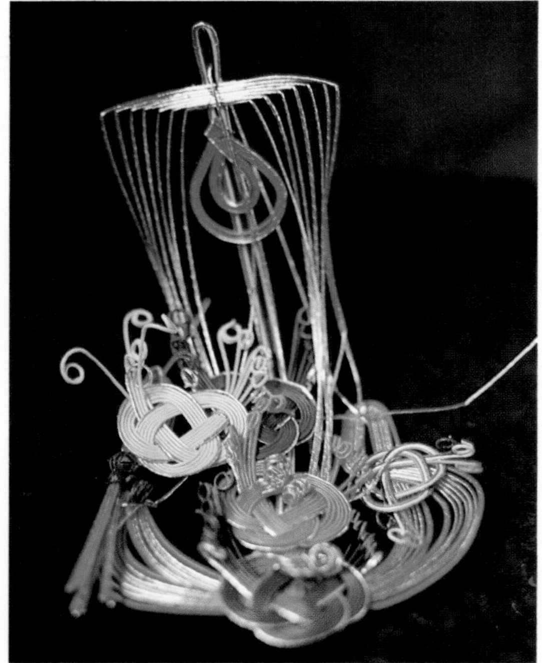

12. Woven and lacquered paper (*koyori*) objects: hat, bottle, lantern top, pouch, cups, ladle, tea caddy.

13, 14. Year-end Daruma market.

15. *Dondo-yaki:* burning last year's Daruma.

16. Antique Miharu dolls (papier-maché).

17. Woven paper fabrics (*shifu*): farmer's coat; warrior's undershirt.

18. Woven paper fabrics (*shifu*): bolt with rough weave; fine wrapping cloth.

19, 20. Paper clothing (*kamiko*): kimono with silk trim; *chabaori* coat; quilt cover.

21. Scroll painting of stencil dyeing, early 17th century.

22. Lacquered and *shibu*-coated objects: temple cushions; hat; boxes; dustpan.

23. Japanese books.

24. Fans: *sensu* and *ōgi*.

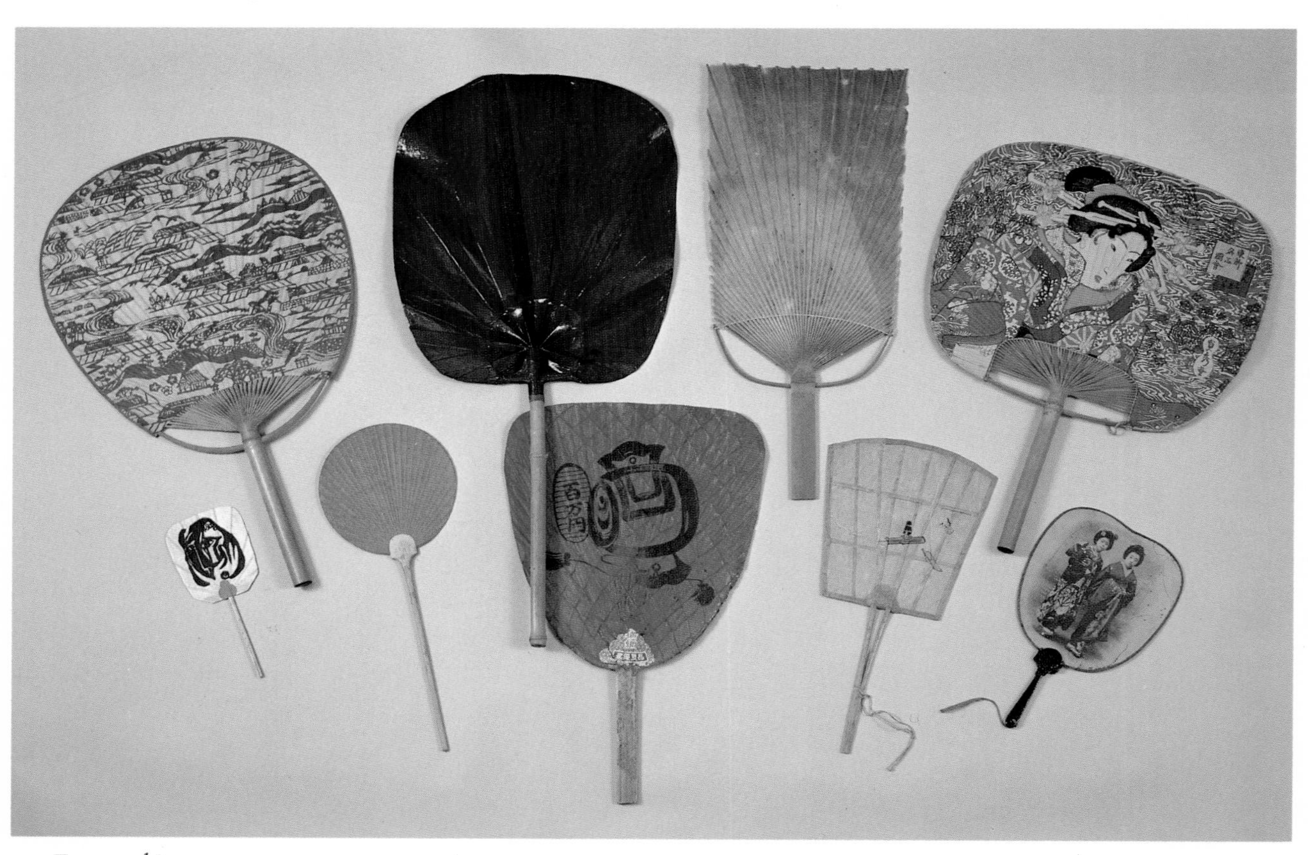

25. Fans: *uchiwa*.

26. *Shōji* with repairs.

27. Eishirō Abe's workshop, Izumo, Shimane Pref.

# On Washi

Among my earliest recollections of childhood in Japan were the many gifts of handmade *hanshi* paper that tradesmen left at New Year's time. The white paper that ushered in the year would be useful throughout in innumerable ways. What would the Japanese household do without this paper? It is the source of light. The shoji screens must be repapered each year or mended. Gifts must be wrapped in it and tied, when important, with the ceremonial red-and-white cord also made of paper. Cakes on any elegant occasion are served on it, as in the tea ceremony. Depending on its grade and availability, paper was used for every conceivable purpose. It has always been a handkerchief. As origami it is the plaything of children.

White is the sign of purity, of birth, and of death in Japan. The use of paper, white, untreated handmade paper, is that of reverence. Folded and cut as streamers and hung from ropes it consecrates all places: shrines, venerable trees and rocks. It encloses the ceremonial areas for combat, as with sumo wrestling for instance, or for ritual creativity, as I have observed in the making of gifts for temples.

Washi may have come from China or Korea as so much else in traditional Japan. But certainly the particular papers discussed in this book made of *kōzo*, *mitsumata*, and *gampi* are characteristically Japanese. By concentrating on the basic fiber and treating it without adulterations, Japan has achieved papers of exceptional purity and strength. The papers of *kōzo*, the inner bark of the paper mulberry tree, is exceptionally so. This is the paper that has permitted the peculiarly Japanese development of the folding paper lantern, the folded paper fan, and the paper umbrella. The latter is treated with waterproofing oils. We find paper used as the foundation of sculptures in the Tempyō period, which are then treated with lacquer or with various pastes of powdered seashell and earth. Then came the marvelously light *byōbu* and *fusuma* (screens and doors) made of layers of paper upon a lattice of wood, which are then painted. All this, of course depends upon the abiding quality of washi, as is appreciated by those today in Italy who restore paintings damaged by the flood of the Arno.

*Washi* means "Japanese paper." *Mingei* is the folk art of Japan, which was uniquely recognized and popularized by that extraordinary trio of Sōetsu Yanagi, Bernard Leach, and Shōji Hamada.

An eloquent section of this book studies the nature of Japanese tradition and the place of the unknown craftsman, which the Mingei movement celebrates. Mingei is, of course, hand manufacture of articles of common use made under other conditions than today's. It survives as a heritage in popular esteem, while its aesthetic, which was a part of its validity, is a challenge as well as a lesson for our times. With respect to washi, I find this best expressed in a remarkable quotation from Yanagi included here.

To capture in words the nature of paper, its relation to nature, the life experience of its maker, and his technical travail is to me a sort of breathless magic. Perhaps it takes a foreigner to perform this—as with Mingei, in a way.

My own first use of the handmade papers of the Orient was in Peking in 1930, when I became involved with "writing flowers" with sumi ink on large sheets of that incomparable paper of China, a rag paper, I believe, similar perhaps to Japanese *gasenshi*. Here the ink absorbs and spreads with that perfect control, which is the envy of the Japanese.

Next came my preoccupation with sculpture as integrally reflected light, then as translucent light, which led me again to my earliest memories of light filtering through shoji screens.

In 1951 I embarked upon my efforts to transform the traditional spiral paper lantern into an object of art, into light sculpture as I conceived it.

I was in Japan; paper and bamboo fit in with my feeling for the sensibility of light. Only washi of *kōzo*, the type called *Mino-gami*, had the strength. With repeated folding or crinkling it only seems to gain in quality, the light casting its shadows like that on the paper-thin gold of a Minoan mask.

The name AKARI, which I coined, in Japanese means light as illumination. It also suggests lightness as opposed to weight. The ideograph combines that of the sun and the moon, exemplifying lightness (as essence) and light (for awareness). Their quality is poetic, ephemeral, and tentative. Looking more fragile than they are, they seem to float, casting their light in passing. They do not encumber our space as mass or as possession, as if they hardly exist in use; when not in use, they fold away in an envelope.

How could I have conceived such possibilities except through a knowledge of the qualities inherent in washi. I hope I may have extended its appreciation through the light and shadows it casts. This is what comes from an old tradition, the use of washi and the making of lanterns, which extends its life into the needs of our time.

Isamu Noguchi

# Preface

When I first began making paper my teacher, Seikichirō Gotō, gently said to me, "Even if a person doesn't make paper very skillfully, if he makes it honestly, it will be good paper. You can see it in the sheets. A craftsman must be honest from the spirit! If he avoids work or menial tasks, he's not sincere about papermaking."

I took his words to heart. Through the winter of 1970 I commuted to his studio at the foot of Mt. Fuji, stoking the pot containing fiber and lye, washing the dusty drying boards and plunging my hands in and out of the cold, cold paper solution as I learned to form sheets of "pure grass" Japanese paper.

By the winter of 1969 I had been living in Japan with husband David for about six months and had begun to write magazine articles on Japanese folkcrafts. One day I saw a television program devoted to the master papermaker Ichibei Iwano. What an extraordinary amount of labor and time he spent producing this stuff, this handmade paper, I thought—labor performed under the sun and wind, sometimes knee deep in a river. And Iwano's rough, artless character impressed me. Suddenly I was so strangely thrilled with the subject, papermaking. I researched a bit and traveled to papermaking sites with photographer Marty (Martha) Cooper. This produced a magazine article on papermaking and plans—cheekily enough for me at this stage—a book!

Our second year in Japan brought us to a school for international Japanese businessmen at the foot of Mt. Fuji, where David taught. Through Mr. Katō, friend of a friend, I discovered that there was a papermaker in town, Seikichirō Gotō, an artist-craftsman and writer. When an introduction was offered, I jumped on the opportunity. I met Mr. Gotō in late October of 1969. Mr. Katō and Mrs. Gotō acted as interpreters. I had only hoped that Mr. Gotō would allow me to follow him around as he went about his papermaking activities that fall, and I would take notes. But somehow in the language shuffle Mr. Gotō thought I wanted to learn the process itself from him, firsthand. He generously said yes, he'd teach me how to make paper. When I realized what the arrangement was to be, I thought—well, why not? It may be interesting.

And so I came into papermaking sideways, as it were. Nearly every weekday morning I arrived at his studio, ready to work and to learn. I knew I was much more of a nuisance to my teacher than a help. From time to time he left me alone to work out the day's chores and their problems for myself. Usually I was waiting for him when he returned, unable to proceed any further in the various difficulties I encountered. But never will I forget my wonder and joy when I peeled my first crisp, imperfect sheets from the drying boards. What a magical thing this creation of paper! How strangely grateful I felt to the fibers for their malleable spirit!

At one point Mr. Gotō felt I had become good enough to help him make paper for his own books, on slightly larger screens; but my sheets were uneven and my awkwardness at removing them from the screen left them frequently wrinkled, or broken at the corners. He never said a word about these sadly formed sheets. I don't know whose disappointment was greater, his or mine.

By the end of spring I was to make my last batch of paper before returning to the United States, and my teacher's first words were still fresh in my mind. I wanted to make paper as entirely on my own as possible, and without shortcuts. I cleaned the fiber, boiled it, and painstakingly picked away the small bits of black and green bark that stubbornly clung to it. Then I beat it with a wooden mallet on the tile floor of the bathroom (much to the distress of those underneath) and washed the beaten pulp over and over again in the bathtub. Then I scooped it into sheets.

The dried, finished sheets were not terribly good—summer had set in with its great humidity, I had worked too slowly, and the pulp had a slight tang of creeping fermentation to it as I formed it into sheets. But it was the best I had made—the most evenly formed and the whitest. Mr. Gotō called it "American washi."

From then on I was hooked. Washi was an object of complete fascination for me, an infatuation often beyond the comprehension of many people. I am still asked seemingly by everyone why I love Japanese handmade paper so much. But how can I explain the complex feelings that have grown and been nurtured within, that stem from my relationship with this wonderful substance and with Japan?

Washi is warm, tender, human, quiet. If made sincerely and without shortcuts, it has an inexpressible purity. One could call it humble, for it never presumes to be other than the simplest of substances. Its gentle nature asks us, tempts us to use it thoughtfully, but also easily. Using it is apt to bring us closer to essential things and to this earth, which is our home.

But my study of washi has meant more than experiencing its special beauty. In discovering the spirit and origins of paper, I discovered a blank page onto which I could read everything I was ready for and open to. It became clear as my research progressed that, as in personal growth, the more one learns, the more one realizes that ways are infinite and paths leading there limitless. This study could take me anywhere I wanted to go. Every small fact I learned about washi was a key to understanding Japan; and every insight I gained about the Japanese mind revealed another dimension to washi's mysteries.

After all is said and done, it is people who make washi and people who use it; and it is the papermaker's unique relationship to his craft that is at the core of washi's fascination. The affection and care with which the Japanese craftsman regards his materials and his tools, with which he uses those household objects made by hand, and with which he performs the rituals of his work all reflect a reverence for the whole of life. They are deep, inner responses rooted in artlessness and familiarity, which seem to thrive in that uncomplicated state of mind, mostly lost to us, where life, work, and beauty are not separate, where spirit permeates every simple activity. To see life and craft touching and unfolding is something very special—a counterpoint of two harmonies, each made richer and more beautiful by their merging. An unarticulated reverence for all existence streams out of these folkcraft traditions; and where this happens, as among these papermakers, we see life itself become a work of art.

## *AUTHOR'S NOTE*

*Washi* is the word for Japanese paper in the Japanese language. *Wa-* as a prefix means Japan(ese); *-shi* and *-kami* (or *-gami*) as suffixes mean paper, any kind of paper, and these suffixes are often interchangeable. *Tesuki washi* means specifically the handmade paper of Japan, although in this book, as in common Japanese usage, "washi" alone is used to refer to Japan's handmade papers.

"Rice paper" is the term the English-speaking West has used for the past fifty years or so to denote all oriental papers, including washi. This has been a terribly misleading name, for most Westerners now think that washi is made from rice, or at least believe that rice plays an important part in its manufacture. Actually rice has almost no place in Japanese papermaking. Sometimes a poor-quality paper is made from rice straw; also rice polish is sometimes added to the vat solution as a filler or to make the paper whiter, but even this is rare today.

The "rice paper" misnomer goes back to 1927, when a British government official mistook Japanese *gampi-shi* for another paper then popularly called "rice paper," which it resembled. Sheets of both were very thin, translucent, and slightly brown in color tone. The original "rice paper" was actually not a true paper at all, but thin sheets cut spirally from the inner pith of the Taiwanese *Fatsia papyrifera* plant—a pith that reminds one of small-pored styrofoam. This tree has also been called the *tsuso* plant, and for some time the "paper" sliced from it was known as "*tsuso* rice paper." As early as 1805, sea captains and China traders brought this "rice paper" back to England and the United States, where it was used as an inner-lining material for hats and for artificial flowers, picture cards, doll cutouts, and watercolors. It was also the first cigarette paper. The sheets painted by Chinese craftsmen with figures, birds, fish, and other designs were especially prized.

In this book, names of Japanese contemporaries are in Western order (given name first, family name last), while names of historical figures are in Japanese order (that is, family name first, given name last).

A common designation used frequently in the text is "Living National Treasure" (*ningen kokuhō*). These "National Treasures" are craftsmen and artists who have received the ultimate award-recognition conferred by the Japanese government for their mastery of a particular traditional skill. Two papermakers and several papermaking groups have been so honored. The award emphasizes the skill, not the individual personality. *Jūyō mukei bunkazai*, "Important Intangible Cultural Asset," designates a traditional skill that some genius of a craftsman has developed, perfected, interpreted, and merged with himself so closely that man and mastery are one. The holders are given a yearly honorarium to insure them of some income and help them pass their skills along to the next generation. A somewhat lesser but still significant national award is the *mukei bunkazai*, "Intangible Cultural Asset." Moreover, prefectures will often confer their own similar awards on local craftsmen of great skill.

I apologize to the reader for the great number of Japanese terms that appear in the text. I have tried to use their English equivalents wherever possible, but there are inevitably instances in which there is no English word for a concept and Japanese provides the more exact vocabulary.

# Acknowledgments

For their help in the preparation of this book I am sincerely indebted to the many people who unstintingly gave their time and talents, and wish to extend to them my deep thanks:

Seikichirō Gotō, my teacher, who gave so generously and patiently of his knowledge and expertise, refusing to take anything in return, providing numerous introductions, and laying the foundation upon which this work could be built; and his wonderful wife and family;

David W. Hughes, my former husband, for his aid in research, his translations, his helpful comments on the manuscript, and enormous encouragement;

Louise Picon Shimizu, for her extensive and polished translations, help in research, and her great enthusiasm for the project; and her husband, Masaharu, who got involved in spite of himself;

Martha Cooper, for her companionship in travel and the enormous time, effort, and talent that went into her fine photographs in this volume;

Reiko Ide Imamura, who gave so generously of her time and effort in research, interpretation during interviews, and translations, letters, and much more;

Hitoshi Hosokawa, my indispensable "aide" in all aspects of research, translation, and interpretation;

Mamoru "Gun" Iwasaki, for his efforts and talents so apparent in his expressive photographs;

Hisashi "Ralph" Matsunobu, for his enthusiastic aid in a million details, my special thanks;

Tarō Ishibashi of the Japanese Foreign Office, for his deep insights into Japanese feeling and spirit;

Neil Seaman, who unknowingly helped in so many ways, especially toward a more refined and focused thought;

Kichirō Hayashi, Professor of Economics, for his comments on economic aspects;

Shin'ichi Yamashita of NHK Radio, who opened up opportunities;

Yasuo Sasaki, for his research on the Kurodani cooperative, translations of poetry-songs, and for opening doors of thought concerning Japanese spirit and aesthetics;

Koichi Katō, for introducing me to my teacher, for his efforts as interpreter, and general support;

All the businessmen-students at IIST for their invaluable help in understanding and interpreting Japanese culture and feeling;

Mr. & Mrs. Fred Dierks and Mr. & Mrs. Jim Keyes, who provided emergency working space so generously;

And finally to my parents, Mr. and Mrs. Norman D. Buehling, for their constant encouragement and support throughout the many years of manuscript preparation.

Special thanks to the following people and institutions for their generous cooperation and technical advice:

Kiyofusa Narita and Hidetarō Satō, successive directors of the Paper Museum, Tokyo;

Makoto Yagihashi, Researcher, Intangible Cultural Properties, Agency for Cultural Affairs;

Hajime Nakamura, Director of the Kurodani Cooperative, Kyoto Prefecture;

John Patton, Professor of Chemical Engineering, Michigan Technological University, Houghton, and his colleagues in the School of Paper Technology; also Professor Davis Hubbard for his introductions to them;

Hoichi Kubota, Sekishū-banshi Preservation Group, Misumi, Shimane Prefecture;

Directors of the Japan Folkcraft Museum (Nihon Mingeikan) in Tokyo.

Additionally I am indebted to the artists and craftsmen who gave time from their work to allow me to talk to them and photograph them:

Nobumitsu Katakura; Shiryū Morita; Michiko Kawaguchi; Sadao Watanabe; Keisuke Serizawa; Hiroshi Haneishi; and Hodaka and Chizuko Yoshida.

Last, but definitely not least, my sincere thanks to the many papermakers themselves who interrupted their work to teach me about papermaking:

The late Ichibei Iwano, Okamoto-mura, Fukui Prefecture;

Eishirō Abe, Yakumo-mura, near Matsue, Shimane Prefecture;

Saichirō Naruko, Ōtsu, Shiga Prefecture;

Tadao Endō, Shiroishi, Miyagi Prefecture;

Kōzō Furuta, Warabi, near Mino, Gifu Prefecture;

The Konbu family of Yoshino, Nara Prefecture;

Mr. Shimizu, the *gasenshi* maker of Misumi, Shimane Prefecture;

The Horiis, the Ishizumis, and all the other wonderful papermakers in Kurodani, Kyoto Prefecture;

And the helpful papermakers at Najio, Hyōgo Prefecture.

To all of them and the countless others who gave me assistance, my sincere thanks.

*A painting, a poem —*
*How paper reveals a man's soul!*

*Dedicated with love and gratitude to*
*my teacher, Seikichirō Gotō,*
*and*
*David W. Hughes*

# Evolution

JAPAN PRODUCES MORE varieties and a greater quantity of handmade paper than any other country of the world. For almost fifteen hundred years the Japanese people have venerated the simplest handmade sheet and have paid paper the highest compliment by using it wisely, lovingly, and resourcefully. *Washi*—the general term for Japanese handmade paper—has influenced Japanese culture itself. Without washi the entire fabric of Eastern history and civilization would have been different. In Japan it has touched every aspect of daily life. The beauty of its surface, its many styles, forms, and uses, and the numerous objects that utilize and are enhanced by its qualities have arisen from the Japanese practical and aesthetic sense. In turn, washi's own beauty helped mold the sensitivity of the people who made and used it. Washi is a material and a medium of expression, yet it is also an expression in itself—it occupies a place where man's inner world and his external statements meet. Made from trees growing wild in meadows and along the hills, washi was man's re-creation of nature—man and nature reflected both upon each other and back upon themselves.

The Japanese have found in their paper a vehicle for the entire range of human expression, from the most noble, religious, and artistic manifestations to the most common, everyday thought. Long ago, people seldom wasted even a small piece of washi, for paper was an extremely precious item. Most people saw little of it. Less affluent correspondents exchanged the same letter sheet back and forth again, each time writing the latest message between the vertical lines of the old one. Then, even when the sheet became completely black with ink, it was still used—the letter writer dipped his brush into water and wrote in intaglio, as it were, lifting the ink from the sheet as he drew the characters. After the surface was completely illegible, the letter was used for such things as lining *fusuma*, the sliding paper doors that divide one room from the next.

Later, when paper was readily available to all, and most farmers made paper as a winter occupation, paper became the poor man's substance. It was also a substance loved by the jaded rich, symbolizing to them a life of simplicity, humility, and harmony for which they longed. Men, women, and children lived in what must have seemed a paper world. Washi was ingeniously formed, bent, cut, folded, twisted, lacquered, woven, and waterproofed with vegetable extracts. It seemed to be capable of substitution for nearly any other material—cloth, rope, straw, leather, and sometimes even glass, bamboo, and wood.

Paper became one of the indispensable materials of everyday life. The clothing and accessories of the humble—jackets, kimonos, sandals, hats, hair ties, wallets, handkerchiefs, belts—all were frequently made of paper. Paper lined the clothing of all classes. Samurai women carried small lacquered paper boxes in the bosom of their kimonos, the tops peeking out demurely from the front folds. Sheets of paper were cut into one continuous strip, twisted into "yarn" on a spindle, and woven into a wonderful cloth called *shifu*. Thick, stiff paper was formed into high hats for priests and nobles and lacquered black. Paper string was woven and molded into fantastic shapes and lacquered to make purses, boxes, luggage, hats, water containers, tobacco cases—the variety of ingenious objects so devised was seemingly endless.

Men and women carried paper umbrellas in snow and rain and paper parasols in the summer sun. Fathers made or bought toys of paper for their children—dolls, animals, warriors, paper rattles, and masks. Paper luck charms filled knicknack shelves and household Shinto altars. Paper kites flew from beach and field to catch the winds, and paper carp, giantly colorful, soared high on poles in spring to celebrate the number of male children a family was lucky enough to possess. At night paper lanterns lit the path and kept the wind from blowing out one's taper. In summer, paper decorations streamed through the evening air as people watched for the union of the two constellations of lovers at the Tanabata festival. And perhaps in the deep of night one could catch sight of a silhouette falling on the paper *shōji* screen as a geisha strummed her shamisen for her night caller, or as a maiko dancer coyly flashed a graceful paper fan. Paper, in its soft translucency, cloaked in faint suggestion the mysteries of such secret and private worlds.

Throughout the country's long history, since the inception of papermaking there, Japan has cultivated the use of paper as if it was a product of nature. Paper lived harmoniously with pottery, handwoven textiles, articles of straw and wood, trees, moss, and blooming flowers. A home was never built without it. As essential as wood, earth, stone, and bamboo, far weaker but more malleable and just as versatile, paper complemented these materials. Paper was a material of the architect as much as of the man of Tea, the painter, the calligrapher, or the student who practiced the shapes of characters on a sheet. Paper was touched daily from the time one opened the windows to the morning to the time when one finished the last meal and last toilette. Paper was the ever-handy companion placed close to the heart to be used to blow one's nose and discard, or to carry a perfect little tea cake to one's mouth. Paper was a personal substance, a substance men wrote poems to and poems on. Life without it was inconceivable and dreadful. Paper was the substance that carried words of importance on the battlefronts, borne by arrows from camp to camp; or the substance bearing secret thoughts sent by messenger from lover to courtesan. Women wept over paper messages, and then wiped away their tears with paper tissues. Paper was purity itself in Shinto worship; it marked the enclosure of sacred areas and signified the living presence of the gods. Paper was at once the most cherished of substances and the most disposable.

Ever aware of the time, labor, and patience that men and women have put into its making, the Japanese have long cherished the handmade sheet. To the West, paper is synonymous with impermanence and disposability, although recent paper shortages are teaching us to be careful of even the commonest machine-made sheets. Washi, on the other hand, has always been regarded with a special reverence in Japan, a reverence that its availability has never shaken. The Eastern respect for handmade papers will at first seem strange to Westerners, for we cannot easily understand how such a simple material could possibly be worthy of esteem. Handmade paper is indeed a humble substance; but then are not beauty and humility all the more reasons to pay an object homage? This is the way the Japanese feel about their handmade paper, washi.

### *Before Paper*

For as long as man has known how to express himself through illustration and written language he has sought suitable surfaces upon which to record his graphic communications. Ivory and bone; wooden boards, sometimes coated with wax; prepared animal hides; tablets of clay; metal plates; the bark of trees ranging from birch bark to the hammered tapa of the South Sea islanders and *amatl* of the Aztecs; woven silk; bamboo tablets; laminated papyrus; stone; palm leaves—all these acted as writing surfaces long before true paper was developed.

Definitions of paper invariably read something like this: paper is an aqueous deposit of isolated vegetable fiber that has been broken down and felted upon a screen. The meeting of certain criteria is generally thought to be necessary for a substance to be true paper: most significant is that the fiber is vegetable; that it is processed until broken down into separate filaments; and that the sheets are formed by some sort of sieving process, usually with a mold bearing a screen, which is submerged into a liquid pulp of these fibers and the excess water allowed to drain through the screen. This describes paper as it is made today and how it always has been made since its inception in China nearly two thousand years ago. By this universally accepted definition, papyrus, tapa, and Mexican amate are not considered true papers, for although made of vegetable fibers, these are not reduced to individual filaments in processing, nor are their pulps sieved through some kind of screen.

The Egyptians developed papyrus as a writing material in the third millennium B.C., if not earlier, long before true paper was invented. First used to construct light skiffs, sails, cloth, and cording, the papyrus plant was eventually employed in making the long, inscribed papyrus rolls that were the books and documents of the ancients. The stem of the plant was cut lengthwise into strips, which were laid closely alongside one another and then covered with more closely placed strips at right angles to the first layer. Soaked in water, then pressed between boards, the layers of crossed fiber adhered together into a sheet. This laminated substance was finally pounded, sun-dried, and rough areas polished smooth with ivory or shell.

Bark cloth and bark paper were known to surprisingly many primitive cultures the world over. Not only the Polynesians and Melanesians made beaten bark material, but also the Javans, the Mayans and the Aztecs of Mexico, the Tlingits of Alaska, and certain Congo peoples of Africa. How substances so similar developed in some corners of the world yet not in others is indeed a mystery, but the methods used in their manufacture are generally the same. Bast or bastlike bark fiber is separated from the stalk, cleaned and washed, and dried in the sun. In the making of amate (anciently *amatl* or *huun*) in Mexico, the fiber, usually a kind of fig or paper mulberry, is washed and then boiled with lime water derived from soaked corn. The dried bark is then laid on a board, cut to size, and beaten flat with a mallet or stone while moist. A binder such as arrowroot or other sticky substance may or may not be used. The narrow pieces are joined by overlapping the edges and beating them until they adhere.

Although bark papers are not true papers, the relationship between tapa and the most common fiber used in making washi is striking. The bark used is the same—the paper mulberry. In fact botanists believe that the paper mulberry came to Japan very long ago from the South Seas. Like many primitive peoples around the world, the ancient Japanese made and wore simply woven garments of bark. In ancient Japanese literary texts, mention is made of using paper mulberry bark for clothing, wrapping, and other purposes. Moreover, the Japanese *tae*, meaning a rough cloth woven of paper mulberry fibers, derives from an earlier word *tape* (the "p" disappearing between vowels in Japanese after about 700 A.D.)—therefore it is linguistically conceivable that the Japanese term is related to the Polynesian *tapa*. Here, however, the relationship ends; true paper, such as washi, and bark paper have nothing more in common than their basic materials, and to some extent the processing of these materials.

*The Invention of Paper*

EXACTLY how true paper came into being is not precisely known, but scholars have formed a good theory. Although old Chinese historical documents carefully and explicitly credit one Ts'ai Lun, a court eunuch, with the invention of paper in central China in 105 A.D., it is certain that neither the idea nor the product was arrived at overnight. The development of paper evolved gradually over a period of time. Certainly the need for a suitable surface upon which to write existed since the inception of ideographic writing in China several thousand years ago. Books made of strung-together bamboo tablets served for a time, but the material was never really practical because the books were bulky and somewhat heavy—libraries took up too great a space. Wood and silk were also employed as writing surfaces. Silk had the advantage of being light and much more compact than the bamboo and wooden books, and since, when sized, the fabric provided an excellent writing surface, a long text was simply rolled into scroll form. But silk was expensive and not readily available to many.

When the brush with its tapered head of hair came into use in the third century B.C., all writing in China underwent a small revolution, effecting not only the standardization of the written language but also the improvement of other writing materials. The first major advance in the development of paper was, it appears, a felting of residual silk floss and sometimes also silk cocoons on a crude mat into thin, tissuelike sheets. Looking and feeling quite like the true paper yet to be invented, this substance constituted a kind of quasipaper produced much more cheaply than silk cloth. Although not made of vegetable fiber, still it was a sieved product. It was referred to in Chinese by the character 紙 (*chih*). Later when true paper was invented, the same character was used for it, but the new substance was sometimes written with the character 帋 as well. Although no old documents specifically state that the earliest *chih* was this quasipaper of residual silk, it is clear from literary references of the day that this was so. Fragments of Chinese silk quasipaper found in a tomb in Shensi dating back to the first or second century B.C. support this theory.

The resourceful eunuch Ts'ai Lun had risen from lowly circumstances through personal ingenuity, intelligence, talent, and political finagling to the position of head of the Chinese Imperial Supply Department during the reign of Emperor Ho in the first century A.D. Part of his responsibilities included overseeing the palace library and satisfying the whims of his intellectually inclined empress, who had an extraordinary fondness for the writing material *chih* and cakes of black ink. Undoubtedly spurred on by opportunity and the desire to please, Ts'ai Lun bent all his efforts for several years toward the development of a better writing surface.

No one can really state to what extent Ts'ai Lun himself was responsible for the invention of true paper, but it was he who first presented this new substance to the imperial court and received credit for its development. This paper consisted entirely of vegetable fibers—macerated fish nets, used hemp, rags, and tree bark, notably mulberry bark. The very first true paper was evidently a combination of all of these fibers, but soon afterwards paper also came to be made entirely of one kind of fiber alone. Ts'ai Lun's paper adopted the name *chih*, that of the old quasipaper. Both *chih* characters were used in reference to this new, true paper.

From 105 A.D. on, the manufacture of paper was pursued with enthusiasm. Methods of manufacture were still fairly unsophisticated, however. Under the reasoning that any fibrous material once softened and beaten was potential paper fiber, many plants were experimented with—rattan, laurel, China grass, bamboo, and rags of flax, hemp, and other fibers, to name a few. The pulpy solution of fiber was probably poured directly over a coarse cloth screen supported by a frame on the ground; the frame was then lifted at an angle to allow drainage and so left to dry in the sun. Scholars presume that the pulp solution must have contained a mucilaginous starch to keep the fibers bonded together, although some early papers lacked this, having a brittle quality. Later in the development of the craft, the screen-mold was lowered into a vat of the pulp solution. The sheets were sized with gypsum and vegetable gelatins and sometimes additionally polished with a smooth stone or a piece of porcelain. In some parts of Indochina and Asia, methods closely resembling these are still practiced in the making of paper.

From China the craft of papermaking spread outward across Asia along the great Silk Road, the route of trade and travel connecting Europe and Asia and the course later followed by Marco Polo. Chinese Turkestan possibly knew the craft as early as the third century. During a war along the borders of Turkestan in 751 A.D., a number of Chinese papermakers were captured and forced to reveal the highly guarded secrets of their profession. Samarkand in time became a flourishing papermaking territory due to the newly acquired knowledge and natural conditions conducive to the craft. Gradually the techniques moved eastward to Damascus, then on to Egypt and Morocco. Europe received knowledge of papermaking quite late, in the twelfth or thirteenth century through Italy and Spain. At first condemned because it was introduced by the then distrusted Arabs and Jews, paper eventually gained acceptance as an indispensable and fairly inexpensive writing substance.

The early European papers were made of linen and cotton rags, just as they are today. The thickness and roughness of the sheets demanded sizing with animal gelatins before they could receive ink applied with the quill pen, the tool used for writing on parchment. It is no accident, then, that Johann Gutenberg devised the printing press as he did, expressly to adapt the type impressions to the hard and impervious surfaces of European sized rag papers.

*The Development of Paper in Japan*

WASHI, the handmade paper of Japan, is manufactured by traditional methods that have changed little through the centuries. In fact, washi seems to typify the old cliché—it is the foreign product brought to Japan, assimilated into her culture, further developed, improved, cleverly modified, then finally given back to the world as a product uniquely Japanese. This seems to be the Japanese genius. Using the inner white bark of certain bast fiber plants, the Japanese took the basic methods brought overseas from China by Koreans, perfected the techniques, and improved upon them to render their paper distinctly Japanese.

A few Japanese scholars contend that paper existed in Japan long before its invention in China; washi, they suggest, originated in these islands, not on the mainland. Events recorded in old and trusted Japanese documents, however, tend to contradict this theory, and more impartial Japanese paper scholars put this view down as irresponsibly nationalistic. Historical evidence indicates that it was the Koreans, particularly Buddhist monks, who first introduced Chinese papers to Japan. In the fifth and sixth centuries Japan had a territorial foothold in the south of Korea; a constant flow of immigrants from these territories and numerous ships running between mainland China and the Japanese coast brought continental culture, arts, and literature to Japan.

The introduction of Buddhism and the widely accepted adoption of the Chinese system of writing took place almost simultaneously in the mid sixth century. It was natural then that Japan, under Empress Suiko, would also desire to know the craft of making paper. In 610 A.D. the Korean priest-physician Donchō (Tam-chi in Korean) brought the techniques to the Japanese imperial attention, as recorded in the *Nihon Shoki*, compiled in 720 A.D. Crown Prince Shōtoku (regent 593–622) had not only a keen and active interest in Buddhism, but foresaw the tremendous implications on his country's culture by the production of this new substance, paper. Encouraging the cultivation of mulberry trees and hemp plants, Shōtoku Taishi also encouraged the nationwide practice of the craft. He oversaw experiments seeking out the best fibers and developing improved techniques.

Although the government in 701 recognized the advantage of keeping the special techniques of papermaking under wraps, it was too late. Knowledge of the craft spread out from the capital in Kyoto to neighboring provinces, so that by the Nara period (708–794) some nine provinces were making paper. By the end of the eighth century the number had swelled to twenty. During the second half of the Nara period as many as 180 different paper names were listed in official documents. The type of paper Donchō demonstrated to the court was probably made of paper mulberry fiber and hemp rag in combination, although it is uncertain. Regardless, hemp fiber was the main substance used in early Japanese papermaking, soon followed by *kōzo*, the paper mulberry; *gampi*, a member of the *Thymelaeaceae* family, and various other fibers used singly and in combination. Gradually the use of *kōzo* and *gampi* replaced hemp as the leading paper fiber after the capital shifted to Heian-kyō (Kyoto) in 794. But the methods employed were still those passed on from China and Korea by Donchō, with some minor improvements and changes.

During the middle of the eighth century, when Japan was emulating everything to which Chinese culture was able or willing to expose her, Buddhist priests held a tight rein over the Nara government. A devastating smallpox epidemic had induced the empress to surround herself with 116 Buddhist priests, hoping to hold the epidemic at bay with these holy agents and rid the country of the demons that caused it all. Soon afterward, in 764, a civil rebellion was suppressed, and Empress Shōtoku decided to express her gratitude. She commanded that prayer-charms (*dhāraṇī*), be printed with woodblocks on one million pieces of white hemp paper and *kōzo* paper. Each was about 6.5 centimeters wide and from 30 to 56 centimeters long, and each was stored in its own individual small pagoda of wood. Even if one considers only the manufacture of the paper for the *dhāraṇī*, the accomplishment was obviously an immense one for that day and age. Most important to the world, however, is the fact that these slips of paper constituted one of the very first examples of text printing on paper, a Korean *dhāraṇī* antedating the Japanese by a few decades.[1]

As Japan entered the Heian period (794–1184), her skill and knowledge of paper technology was unrivaled in the world. Each region of the country came to be known for its own special types of paper. Beautifully dyed and elaborately decorated papers from this period, still in excellent condition, today can be found in museum displays and private collections. Among the nobility, Buddhist clergy, and government bureaus, there was a fantastic demand for paper. The flowering of literary pursuits at court; the increasing growth and influence of Buddhism, which encouraged sutra-copying as a path to salvation; and finally the machinery of bureaucratic government, which, with the political reorganization in 646, drew up an edict demanding nationwide census registration and taxation records—all these made the production of many kinds of paper in significant volume essential. The oldest extant Japanese paper is in fact a series of census registration documents from Mino, Chikuzen, and Hizen provinces dated 702. Moreover, many papermaking districts would send tributes of paper to the central government, demonstrating their loyalty and hoping in return to receive protection in the form of soldiers during these war-ravaged years.

In 807 the government in Kyoto set up a paper mill there called the Kanya-in (or Kamiya-in). Its expert papermakers had been brought in from the country mills to meet the needs of the court and governmental agencies for quality paper. Competition within the mill among craftsmen led to improved techniques and innovations. The method of *nagashi-zuki*, a unique method of handling the solution on the screen and found only in Japan, was thought to have been developed at the Kanya-in during this period. Kanya paper rapidly became the finest available, much sought after by the well-born and wealthy. This government mill annually produced some twenty thousand sheets of fine white paper for imperial use. Evidently the Kanya paper far surpassed that still imported from China and Korea, for, as Lady Murasaki Shikibu wrote in her famed *Tale of Genji*: "This Chinese paper is brittle! I fear it would fall apart if handled from morning to night. Send for the workers of the Kanya-in and have some paper specially made."[2]

By the mid-Heian period, the central government in Kyoto had lost its firm grip over the outlying regions, and the provinces stopped sending fiber material to the Kanya-in as tribute. The jealously guarded Kanya-in papermakers, who were practically government slaves, began to escape to the countryside, where they joined the staffs of mills owned by private estates. Deprived of raw materials and good craftsmen, it was only a matter of time before the once-great Kanya-in mill fell into decline. Meanwhile, papermaking in the provinces gained new strength and vigor.

In the latter days of the Heian period, the Kanya-in turned in desperation for its survival to what the Japanese came to consider an unclean, inferior method of making paper—recycling it. This paper of dark gray color, tinted by the ink of used sheets, was called *suki-gaeshi*, or *shukushi*. Originally *shukushi* had been a fairly noble type of paper. Aristocrats, particularly the more sentimental among them, sent letters, poems, and sutras written by their beloved deceased to the papermakers to be remade into "new" paper. This remade paper was particularly loved for its soft, light gray tone, which resulted when the ink on the used paper dissolved in the pulp solution. On these sheets those of gentle birth copied sutras as an act of piety. Eventually, however, as the quality of its paper declined, *shukushi* became the Kanya-in's sole product, and the raw materials now were whatever used, dirty, and damaged sheets could be collected. Recycled paper had lost its glamor.

As the Heian period drew to a close, paper made in outlying districts came to rival in quality that made at the height of the Kanya-in's glory. The provinces were active, vital, and increasingly wealthy. Better roads brought capital and countryside closer together. Papermaking was fast becoming an important cottage industry—as soon as a new type of paper was developed, the people seemed to be able to find an urgent use for it. Beautifully colored and patterned papers were sought after for poetry writing; such papers were also gifts among the wealthy and the noble. In the provinces paper was a tribute of tax, as negotiable as rice, cloth, and metal currency. *Danshi*, a richly thick, white, and textured paper became the official court paper in the early 1300s and remained so for over five hundred years (see p. 206).

During the Kamakura period (1185–1333), political power shifted from the imperial capital in Kyoto to Kamakura, where the military aristocracy had its seat of government. Feudal law was codified. The provinces experienced a greater refinement in and higher standard of life. Simultaneously, Japan experienced a religious awakening. Zen Buddhism became the core of samurai activity and thought. Art and literature thrived in the Zen monasteries. Asceticism reigned. Paper garments were stylishly plain, humble, and impoverished in spirit.

Around 1330, at the beginning of the Muromachi period, Japan experienced tremendous economic growth. Estates, which had been little economic units in themselves, began to be amalgamated into large fiefs. Trade expanded enormously, aided by the establishment of a systematized money exchange. Markets developed. Like other handcraft and agricultural products, paper was now produced extensively for trade as well as for local consumption. The restrictions imposed by the feudal structure of society led to the formation of guilds, both for craftsmen and specialized merchants. These guilds, or *za*, linked papermakers with certain noble families who acted as their patrons, affording protection in the monopoly of highly specialized papers. Not infrequently influential families smuggled into their region papermakers from other parts of the country to teach secret paper techniques. The guild system was an effective one and encouraged the growth and development of papermaking throughout the provinces, particularly through a flourishing fifteenth century.

The first paper money appeared, reportedly, around 1620, when certain merchants in Ise issued paper notes as change for the long, oval-shaped silver coins, when the government forbade the populace to cut the largish coins up to make smaller currency. Other merchants followed suit, imitating the Ise businessmen; eventually even the feudal lords were printing their own bank notes with wood blocks, until the currency situation was so confused that in the late nineteenth century the Meiji government had to universalize one system.

During the Edo period (1615–1868), paper was the second greatest source of tax income for the Japanese government—the greatest source was rice. At this time the provinces felt the necessity to produce paper in amounts sufficient to help support their own economies. In fact, they could not afford *not* to produce paper. Dozens of provinces made excellent papers; hundreds of varieties were available, with hundreds of products made of each variety. Echizen, Yamato, Mutsu, Mino, Bitchū, Satsuma, Iwami, Settsu, Izumo, Hitachi, Iyo, Chikugo, and Suō were just a few of the best-known and established papermaking areas at this time. Roads and flow of commerce during Edo times were such that every region had its area markets, and sooner or later all the various kinds of papers were available in Kyoto or Osaka. So much paper was made and some of it so cheaply, such as *hanshi* (see p. 176), that for the first time in history paper was readily available and accessible to all classes of people.

One colorful outgrowth of provincial papermaking was the Ni-Shichi Fair held throughout the Edo period in the castle town of Shiroishi, near Sendai. Shiroishi was a major paper-producing center in northeastern Japan. In his book *Shiroishi-gami*, Shinichi Kannō beautifully describes this medieval fair. *Ni-shichi*, which means "two-seven," marked the 2nd, 7th, 12th, 22nd, and 27th days of each month, when the market-fair was held. During the dead of winter, when the farmers had some spare time away from their fields, they would attend the paper mulberry stalls at one end of the fair to purchase *kōzo* fiber. Working hard into the night for several days, they processed the fiber they had just bought, forming it into sheets of paper. On the next Ni-Shichi day they would wrap their paper goods in thick paper that had been waterproofed with persimmon tannin and, shouldering the bundles, take them into town. Buyers came to Shiroishi on these days from all the outlying districts, carrying a set of scales and their lunches, ready for the bargaining and hustling of goods and skills.

From papermaking's earliest inception in Japan, the greatest statesmen actively encouraged its development. When Tokugawa Ieyasu took command in the early 1600s and strengthened his hold on an already unified nation under an Edo (present Tokyo) shogunate, the feudal lords put every effort into furthering economic advancement in their fiefs. Economic advancement naturally implied further progress in paper production, progress in quality as well as quantity. Paper was not only an important and valuable commodity used in every aspect of Japanese life from the house to the temple and shrine, paper was a symbol of culture as the cardinal vehicle for the written word.

By the seventeenth century, paper was already being presented as a representative national product to the heads of nations with whom Japan conducted foreign trade. Japanese fans were always among the gifts bestowed upon overseas sovereigns. The Jesuit missionaries that came to Japan were especially admiring of Japanese papers. They used them to print their books in Japan and to a large extent were responsible for much of the washi that managed to reach Europe at the time. The Dutch, however, were the foremost purveyors of washi to the Western world. As the only Europeans granted trading rights in closed-door Japan, the Dutch were instrumental in opening up Europe's eyes to Japanese paper. Rembrandt, for one, was not only aware of washi, he used it frequently for his sketches and etchings. Bunshō Jugaku wrote in his *Paper-Making by Hand in Japan*, "Those who visit the Rijksmuseum in Amsterdam today, will find there some Rembrandts etched out on Japanese paper, and be impressed by the good taste with which he chose only the best papers for his purposes."[3]

Prosaic as paper tissue may be to us now, the use of soft handmade paper by the Japanese as handkerchiefs completely fascinated and enthralled Europeans of the early 1600s. The custom of using paper to clean the hands and mouth goes back to sixth century China. Borrowing this genteel habit from their cultivated neighbors, the Japanese, both women and men, for centuries have carried folded pieces of soft, white paper in small packets kept in their sleeves or tucked into the front fold of the kimono. It served as Tea ceremony napkins, tissues, and emergency toilet paper. Originally a type of *kaishi* (see p. 179), such paper received the euphemism *hanagami* ("flower paper"), presumably because it was discarded like the rain of cherry blossom petals. Because of one of its more common uses, and because the word for flower has a convenient homonym—written with a different Chinese character—*hanagami* today means "nose paper."

At the beginning of the seventeenth century, Lord Date Masamune sent an envoy to the Pope in Rome. The *hanagami* carried and continually discarded by envoy Hasekura Rokuemon caused quite a stir among Europeans, who had never seen paper handkerchiefs. Quoted here from the book *Shiroishi-gami* is the account of Hasekura's visit to France, an account originally written by the Marchioness of St. Tropez, and subsequently retold and rendered into very creative English by an unknown translator:

> The Marchioness de St. Troppez [*sic*] mentions in her description of the gorgeous custom among the Japanese that was much talked about among the inhabitants of the town St. Troppez. It was about the handkerchief used by the Japanese. It greatly differed from the square cloth used for the same purpose in France in the beginning of the 17th century. They blew the nose with a handkerchief made of Chinese silk paper about the size of a man's hand and every time they blew the nose they discarded the paper on the ground never to use it again. The Japanese treasured up the paper handkerchiefs in the bosom as if they had been gorgeous love letters from a court lady. They had a considerable amount of them in the bosom and had provided for a sufficient amount to satisfy their need on their long journey. The inhabitants of the town Sainet [*sic*] Troppez were said to have waited for the Japanese to stir out and no wonder. When any one of the Japanese used the paper handkerchief and threw it away on the street they would run up to him to pick it up. They even fought among themselves in order to secure one of these priceless momentoes [*sic*]. It was the paper handkerchief used by Hasekura that was particularly regarded as being most precious. No doubt Hasekura's paper handkerchief had a high historic value to the public eye. Having been amused at the people's freakishness, the envoy seemed to have used more paper handkerchiefs than was actually necessary so as to please the people of the town.[4]

Of course the paper handkerchiefs were more probably Japanese, not Chinese, as the story states. Since Date Masamune was lord of Sendai, the *hanagami* was probably a product of the excellent local Shiroishi papermakers.

As the seventeenth century came to a close, the German scientist Engelbert Kaempfer traveled to Japan with a Dutch trade mission and managed to persuade authorities there to allow him to observe and record, among other wonders, Japan's papermaking techniques. His observations were published in his *Amoenitatum Exoticarum* of 1712. His edited manuscripts were published posthumously in English in 1728 under the title *The History of Japan*. In his writings Kaempfer presented an extraordinarily accurate—especially for that time by a foreigner—description of Japanese papermaking, thereby giving Europe its first comprehensive view of washi manufacture.

Two concurrent and interrelated events were to have a dramatic and crippling influence on the papermaking industry in Japan—the Meiji Restoration and the development of machine pulp paper. The effect of both fell hard on Japan in the late nineteenth century. With the coming of Commodore Perry to Japanese shores, forcing this closed nation to open her doors to trade and communication with the rest of the world, Japan entered a period of reassessment. The Meiji Restoration (1868) was a necessary attempt to catch up with the technology of the West. Its ramifications included setting up machine pulp paper mills and attempting to industrialize paper production.

The small hand paper mills survived exceptionally well, despite ever-increasing production of machine-made paper, until after the Russo-Japanese War in 1905. The demand for paper was too great to be met by the hand mills, and hand papermaking steadily lost the fight as craftsmen tried to compete with the low prices of machine mills. Naturally the hand papermaking industry suffered severe setbacks. Hand paper craftsmen attempted to modify or semimechanize their methods. Most of these "improvements" damaged washi quality severely. Harsh chemicals, often in excessive amounts, were cooked with the fiber to expedite its breakdown. Cheap wood pulp was added to good bark fiber. Washi suffered.

Some mills succeeded more or less in their battle to survive. Others were forced to close down, and the paper craftsmen took up other work. What is most surprising is the number of hand mills that did survive and kept the old traditions alive and pure, to pass them on to future generations. The cottage mills that still produce the many varieties and colors and variegated complexions of handmade paper—these, I think, are a tribute to an extraordinary way of life, to a unique culture with a set of values that postulates that things of excellence shall not die.

### *The Sacred, the Magical, and the Mundane*

IN JAPAN, much has been made among those who love paper of the fact that the word for god (*kami*; 神) has the same pronunciation as the word for paper (*kami*; 紙). Linguistically there is probably no relation whatsoever between these homonyms; their written characters as far as is known have completely different etymologies. But for the Japanese there is often a kind of connection between these words. Ever since papermaking has been practiced in Japan, plain white paper has been a symbol of purity and godliness. To this day it is used extensively in Shinto ritual and worship. Some feel that white paper represents the very essence of Shinto. To a lesser extent white paper is also symbolically used in Buddhism, as well as in superstition and magic. These uses, however, are most likely an offshoot of cultural associations that link ideas of purity with whiteness and with paper, associations that find their origins in ancient Shinto symbolism, now deeply ingrained in the Japanese consciousness.

A love of cleanliness is innate in the Japanese, Kakuzō Okakura once wrote. In fact, the associations of cleanliness and purity with whiteness must be considered archetypal, for in cultures the world over this equation is universally the same. Why is this? It is probably because a white surface is the most easily defiled. A refined white surface is also difficult for a technically unsophisticated culture to produce—and once produced, or found in nature, equally difficult to maintain in its state of virgin whiteness. To the Japanese mind, white is pure because it immediately betrays every blemish and stain.

The inner white fibers of hemp (*asa*) and paper mulberry (*kajinoki* or *kōzo*) were anciently used to weave textiles in Japan. People fabricated clothing of these kinds of cloth, which were important items in the households of new brides. Undyed hemp, especially, came to symbolize womanly virtues still greatly stressed in Japan—faithfulness, chastity, and obedience. Like the white cloth, a woman, an old saying goes, must allow herself to be dyed any color her husband chooses.

Shinto is more than a formal religion in the Western sense; it is a total worldview and a way of regarding man's cosmic purpose on earth. An outgrowth of animistic folk faith and shamanism, Shinto centers its beliefs around spirits or gods (*kami*) who dwell in rocks, trees, mountains, or other natural phenomena. These mutable spirits are dual-natured, at one moment affectionately protective while at the next fiercely punitive and terrifying. In Shinto, man is a continually evolving creature, ever striving for purity. Although he carries on a passionately emotional, personal dialogue with the *kami*, the worshiper seeks and receives through the gods a spirituality that binds him to the community in a social and spiritual unity.

Purification and cleanliness are at the core of Shinto. Man must continually seek to refresh his spirit, to cleanse his body, his mind, his home, his altar and shrine, his clothing. All that is clean, pure, and honest, all that is bright, chastely natural, and happy display the essential virtues of Shinto.

Hemp and mulberry fiber as well as the cloth woven from them, natural and undyed, joined rice, salt, and saké as offerings to Shinto shrines. Strips of these fabrics were also tied to branches of the sacred *sakaki* tree as offerings. When hemp and mulberry papers were first made, worshipers extended the symbolism and substituted paper for the cloth and bark fibers. Today this particular rite is still practiced with strips of folded white washi tied to *sakaki* branches.

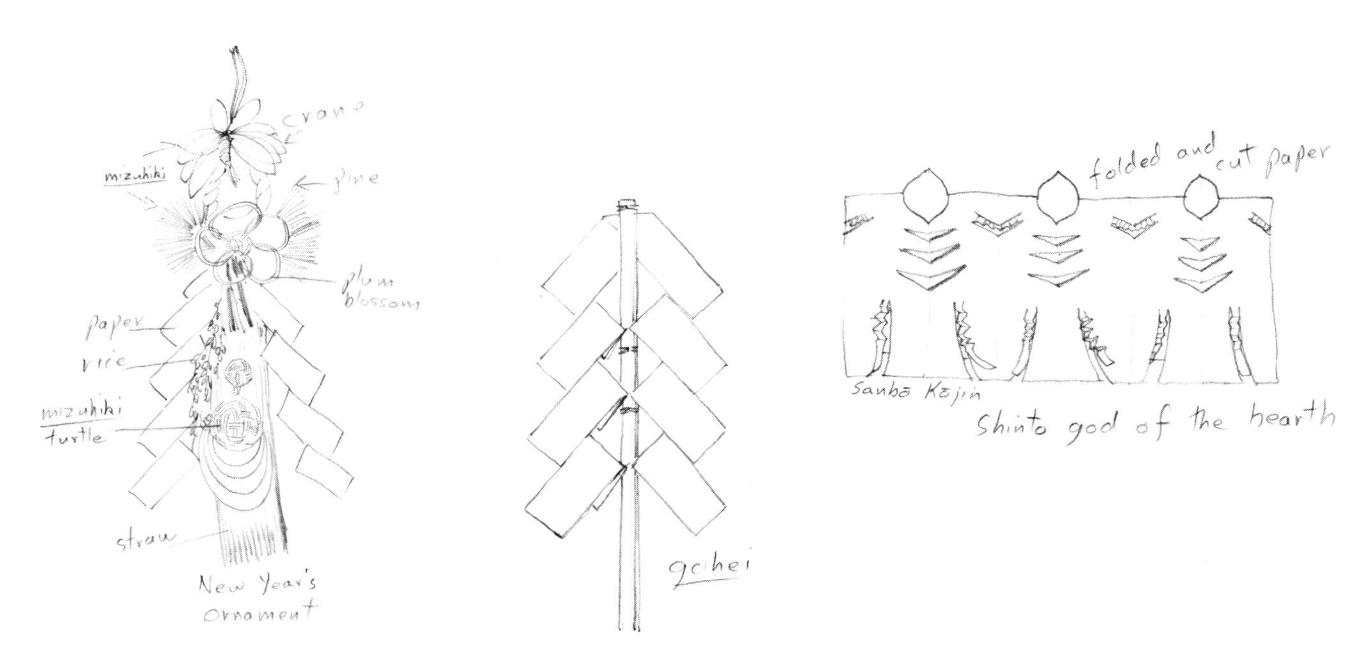


Anciently Shinto devotees believed that their *kami* moved from place to place. To beguile them out of their sacred groves and into the shrines, the people marked off areas of nature that were unspoiled by man's touch; they bound the area off with straw rope and hung upon it pieces of hemp or flax fiber or cloth. Later, these were replaced with ritually folded white paper strips placed at set intervals, called *shide*. Thus the *shimenawa*, the rope with sacred paper strips, symbolically marked the division between sacred and profane places. All shrines today are marked off with these beautiful, starkly simple "taboo ropes," which guard against the entry of evil spirits. They also encircle sacred trees and rocks, barns and purified building sites—in fact any object or place that is sanctified.

Sumo wrestling, the oldest Japanese sport, is an offshoot of Shinto worship. Sumo contests took place at public festivals, where they served as a means of divining the will of the gods and as an act of supplication and communal prayer. Today sumo wrestlers wear hanging from their belts cut *shide* of silk, which, it is said, were once made from the ceremonial paper, *danshi* (see p. 206). In the Nō theater, some of whose origins are also in Shinto, similar paper strips are hung high to sanctify the event and as a symbol of the exorcism of evil spirits.

Folded strips of pure white washi are also attached to a kind of wand called *gohei*, which is employed in Shinto purification rites. *Gohei* can mean either just the paper strips folded like *shide* that hang from both sides of the stick in zigzag fashion or the entire wand. The *gohei* has been called by Munemichi Yanagi (Sōetsu's son) "the most Japanese of symbols for a deity."[5] Altars in home shrines bear these wands, where different forms of cut and folded paper stand for the spirits of wind, water, and fire. The stick of the *gohei* was originally the stem of a young paper mulberry, and the paper strips were white hemp fiber. First an offering to the gods, gradually these abstractly patterned paper strands came to represent the gods themselves or to indicate their presence. In purification ceremonies, Shinto priests will shake *gohei*, often over objects or people, to exorcise evil spirits. In earlier days of Shinto, the devout took on their long

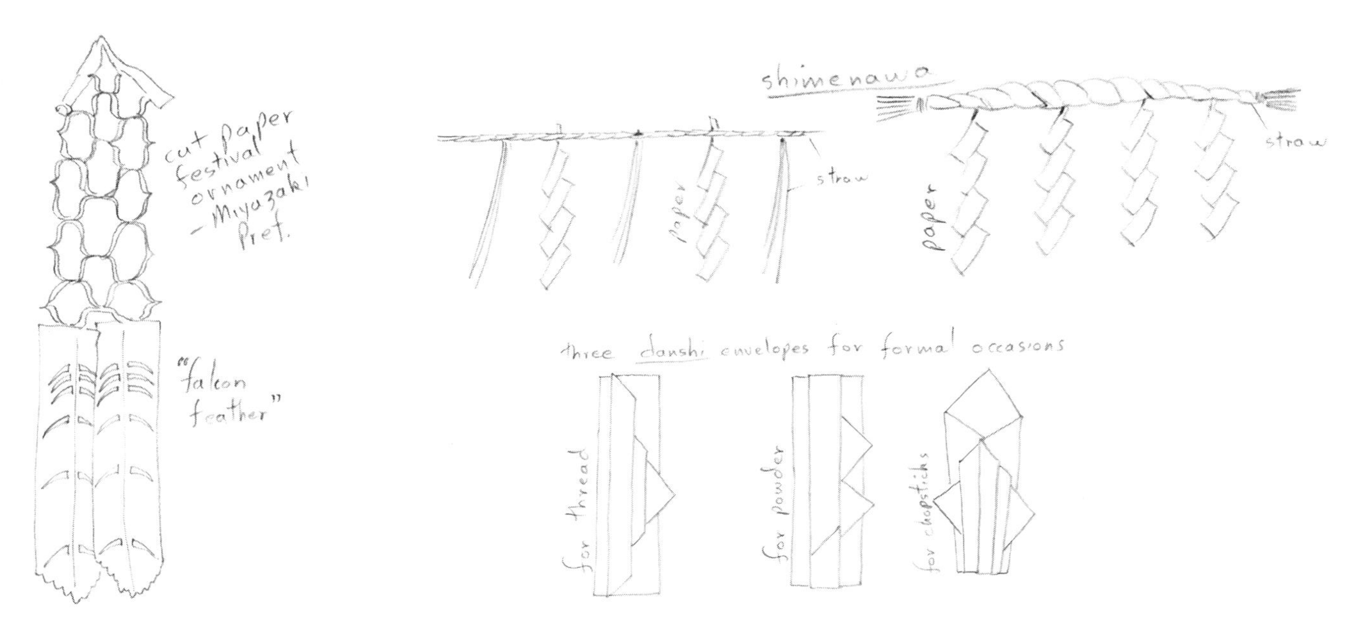


journeys small pieces of cut paper called *shohei*, which they scattered on the road before them for divine protection.

Ritually folded paper as used in shrines and for the wrapping of ceremonial gifts is quite old in Japan. Today's *origami*, the charming folded paper toy structures, is related to the most ancient and serious ceremonial folding of paper, such as the making of *shide*. The ceremonial folding of washi became extremely elaborate, and since Shinto lacks the icons of most other religions, the complex cutting and folding also became highly symbolic.

Anyone who has become acquainted with Japanese knows that they are great gift givers. Few know, however, that there is an entire etiquette governing the proper presentation of the most trivial of gifts. A saying from the twelfth century goes, "When you send someone a gift, always wrap it in pure white paper." This custom goes back to very early times when rice was carefully folded in pure white paper to be presented as offerings to the shrines. The corners of the package were twisted together in a special manner to prevent the rice from falling out. Coins as well were later offered to the shrine wrapped in white paper. It is interesting that a type of currency commonly distributed in old Japan was coins sealed in washi envelopes with the signet of the bank and duly marked for value; the packets were passed from hand to hand, unopened. Even today when money is passed from one person to another, either as a gift or in a business transaction, it is polite to wrap it first in paper. Special envelopes are now made for this purpose.

In Japan, paper fortunes are often bought or obtained without charge from shrines. Out of a large box, the hopeful chooses a bamboo stick, the hidden number on it leading to a slip of paper bearing either a good or bad fortune. Some Japanese believe ardently in these messages thought to come from god-spirits, and they guide their lives by them. Another paper talisman popular in Japan is what is called a "dream-treasure poem." According to custom, the first dream of the new year is highly symbolic of events fated to come during the course of the year. In order to enhance the possibilities for auspicious

dreams, people will buy these dream-treasure poems from a street hawker. In Japanese the poems read the same backwards and forwards: "In the sleep of the long night I hear, half asleep, the sound of a ship coming in, riding over the waves—oh, what a pleasant dream."[6] The belief is that, by reading the poem before sleeping, one might dream of the treasure ship that bears the seven gods of wealth and happiness—the most propitious omen for the coming year.

Cut-out paper dolls are not usually considered toys in Japan. On the contrary, paper cutouts have for hundreds of years acted as effigies and charms against disease and evil spirits in Shinto, Buddhism, and rural shamanism. In China, paper effigies of men and animals burned along with the corpse were believed to play an important role in helping the dead reach paradise in comfort and luxury. Some people believe that the symmetry of paper-cuts has a protective effect; this is true in such diverse parts of the world as Mexico, Poland, and Japan.

Early in Japanese history, anthropomorphic images made from grass and wood were used in rituals to absorb the sins and bad fortunes of people. The images were then thrown into the river or ocean, carrying away with them the effective evil. Similar exorcism rites using paper effigies were also practiced in China before the fifth century. In Japan, paper cutouts in the shape of kimonos, called *kata-shiro*, are objects used in certain Shinto purification ceremonies that take place in the early spring. The *kata-shiro* and the rites associated with it have survived unchanged through the centuries. The worshiper buys a fresh white paper image at his shrine. At home he inscribes on it his name, sex, and birth date, and possibly also a prayer. Breathing on the paper, he rubs it over his body and returns it to the shrine, which arranges to cast the collected *kata-shiro* into a flowing body of water.

In early spring, in various areas of Japan people make dolls of paper—sometimes a group, sometimes a single couple—and float them down the river in small straw boats. These usually simple dolls represent all the spirits of evil that bother children. Originally the dolls were magical images that the whole family rubbed over their bodies or that took on voluntarily all evil and misfortune. These are regional variations of the custom that gradually evolved into the girls' Doll Festival in March, which still bears some vestiges of the original magical rites.

Paper charms were also used to keep away sickness and pestilence. *Hōkō-san* is the title for a domestic servant. On Shikoku a crude Hōkō-san doll of papier-maché is given to a sick child to hold; then the doll is thrown into the ocean, taking with it, it is believed, the child's illness. The history of the Hōkō-san doll is a touching one. Once a homely but tender-hearted maid by the name of Omaki was in the service of a local samurai. One day the samurai's daughter became very ill with a contagious disease—the doctors announced that they could not cure her. Omaki felt that if she could only catch the disease, perhaps it would restore the little girl to health. And so she exposed herself to it boldly while attending to her little charge, until finally she caught it. Because the disease was so contagious, Omaki was put out upon a secluded island, where she died. Ever since, the Hōkō-san doll has been a charm against children's sicknesses.

Washi is a significant material in Buddhist rite and custom, although it is not as symbolically intimate to major Buddhist tenets as it is to Shinto. Religious devotion demanded, in some Buddhist sects, the painstaking copying of sutras. Paper, of course, was an essential sutra-writing material, and as such received the reverential care and respect paid to all articles of devotion. To a smaller extent this is still true today. The act of copying sutras is considered itself a kind of meditation, in which the self is given a back seat and the more instinctual mind solemnly copies the characters. Many beautiful old copied sutras survive from centuries of these Buddhist devotions. This act of copying must sometimes have been accompanied by great religious fervor, for some sutras were written in blood. Copied sutras were usually presented to temples or shrines. Among the most opulent and finest examples of embellished papermaking in the world are the sutras known as the Heike Nōkyō, copied in the twelfth century by members of the Heike (Taira family) and preserved in the famous Itsukushima Shrine near Hiroshima.

### *Kamiko*

ACTUALLY washi was used early in Buddhism in a more intimate manner than one might suspect, for it is said that the first paper clothing, *kamiko*, was worn by priests as underwear. Whether or not one can trace its origins to such elevated beginnings, it is known that in the eleventh century, when Esoteric Buddhism was being actively propagated in Japan, monks began wearing white paper robes made by themselves in keeping with a way of life both plain and severely ascetic—significantly, the handmade clothing expressed their independence from the aid of feminine hands.

*Kamiko* is nothing more than thick paper specially treated with starch from a type of arum root (*konnyaku*; devil's tongue) for strength and water-resistance. The treated sheets are rubbed and wrinkled until supple. Edges of the sheets are glued together to make one long roll from which patterns are cut and sewn into clothing just as if it was a bolt of cloth. Because it is strong, retains heat well, and is water-, wind-, and damp-proof, robes, coats, jackets, and vests of *kamiko* have been popular for centuries.

In time *kamiko* came to be manufactured all over Japan, especially in areas where paper was made. According to old documents, the towns of Shiroishi in Miyagi, Aki-kawa in Shizuoka, Kei in Wakayama, Yatsushiro in Kumamoto, and what is present-day Osaka were all noted for their production of *kamiko*. Farmers and commoners made their own clothing and bedding from it, since paper was cheap and *kamiko* was easily manufactured at home. Twisted *kamiko* strips were used to secure obi sashes, as thread for prayer beads, and as sandal thongs.

Nothing was simpler, plainer, or humbler than *kamiko* clothing. For that reason it appealed strongly to the samurai temperament and fit the warrior spirit (*bushidō*), which took root and blossomed through the Kamakura and Muromachi periods. *Kamiko* tangibly expressed that quality of ingenuousness toward which the noblest minds strove during Japan's middle ages.

Whereas the poor wore paper clothing because they could not afford anything else, artists, high priests, the nobility, and military leaders became inordinately fond of *kamiko*. That of the poor was rough and rust colored from persimmon tannin, but the *kamiko* of the well-to-do was a pure white, often embossed with patterns or dyed in a variety of intricate colors. A *kamiko* robe about four hundred years old worn by the famous general Uesugi Kenshin is still in existence. Such paper robes were often lined with silk and gilded around the collar and sleeves—*kamiko* wearers of old knew that the prosaicness of the paper made the opulence of the gilt all the more apparent.

During the early Edo period (1615–1868), *kamiko* became more widely worn, one reason being that washi production was at its height and a great deal of good hand-made paper was available. The poet Bashō loved and wore *kamiko*. The writer Ihara Saikaku used the expression *kamiko rōnin* in the late 1600s to describe a *rōnin* (a masterless samurai) so poor he had but one paper kimono—yet would not accept charity.

Now *kamiko* was popular both for inner and outer garments, especially in the form of short coats with silk or cotton linings. Cotton or silk floss was usually packed between

the paper and cloth layers for added warmth. Generally *kamiko* garments were sewn with hemp thread; small triangular-shaped pieces of cloth were used to patch rips. After the middle of the Edo period, *kamiko* went out of vogue, again becoming clothing only for the poor. *Kamiko* linings for garments were used ubiquitously in Japan throughout the Edo period and up until this century.

*Kamiko* robes continue to be the clothing of certain of the priesthood up to the present. Vestiges of this ancient use of *kamiko* can be seen today in the Omizutori ceremony at Tōdai-ji temple in Nara. *Omizutori* means "sacred water drawing." This ceremony is at least twelve hundred years old, dating back to early Buddhist days in Japan. Preceding the advent of spring, Buddhist monks spend two weeks or more in heavy meditation, self-reflection, and ascetic training. During this period, each makes for himself a kimono out of undyed, handmade paper. The monk is required to rub, soften, and waterproof the material for his own robe. Between February and March, at the coldest point of winter, the monks go into seclusion wearing only these robes, enduring austerities in the cold and snow. Smoke from burning incense sticks gradually blackens the *kamiko* garments; holes are worn into them where the monks' knees have rubbed through the paper in prayer and meditation. Then when all the austerities have been completed, the priests return for the ceremony. At midnight the *kamiko*-robed monks proceed down the temple corridors, carrying lighted torches and chanting prayers. In an open area of the compound the monks brandish these immense burning torches, making great circles of flaming light in the night air. Since the fire is believed by these Buddhists to be power against evil, worshipers pursue the flying sparks, certain that such magical flames burn nothing on which they land. Then the sacred water, believed to have flowed all the way from Wakasa in the north to Nara, is drawn from the well. At the end of the ceremonies the monks burn their white paper robes in a purging bonfire. Although the ceremony is Buddhist, it bears touches of Shinto in the sacred use of white paper—especially washi that has touched the body—in this rite of purification.

Today Kurodani near Kyoto and Shiroishi near Sendai, once famous for their *kamiko*, still produce fine-quality *kamiko* on a limited scale. In some very rural parts of Japan, farmers manufacture *kamiko* for their own use just as their parents and ancestors did before them.

*Shifu*

ONE other paper textile besides *kamiko* developed in old Japan, a much more sophisticated product that took considerable ingenuity to develop. This is *shifu*, woven paper cloth. Strips of paper are finely twisted or spun into a continuous thread. This surprisingly strong and supple "yarn" is used as the weft threads (usually) in weaving bolts of cloth. Like *kamiko*, *shifu*, too, is cut and sewn into clothing and accessories.

While *kamiko* is excellent as winter wear, *shifu* is just as excellent for summer wear. It is light and porous, having the unique ability to absorb body perspiration and throw it off rather than retain the moisture and cling to the skin. *Shifu* has served as under-clothing, as summer kimonos, obis, book covers, bags and purses, and notably as finely meshed mosquito netting.

How such an unusual product as *shifu* ever developed is not clear, but sometime around 1650 techniques for making it appeared in Japan. The earliest documented *shifu* production took place in Shirakawa, a town near Mt. Masu in that papermaking district of Fukushima Prefecture. It is just as probable, however, that the first *shifu* was made in Shiroishi, Miyagi Prefecture, since the best *shifu* in the largest quantities issued from this castle town under Lord Katakura Kojūrō, who paid tribute to the Date clan in nearby Sendai.

The idea behind *shifu* very likely evolved out of the ancient use of *kōzo* bark for weaving cloth coupled with the more recent development of strong paper cord. A certain local Shiroishi samurai named Sanada devised a method of making extremely tough cord out of twisted paper for armor bindings. Reportedly, no sword could cut through it. Someone began plaiting this same paper cord into sandals, and discovered that they outlasted two or three similar pairs made of straw. Fire fighters adopted them, soaking the sandals in water before conflagrations to make them impervious to flames and nails. Eventually, someone thought of weaving the finer grade of this paper cord into cloth and discovered that it was a lightweight, soft, and airy material.

Whereas *kamiko* had long been manufactured by the lower-ranking samurai, now *shifu* became the exclusive product of samurai families during idle periods of peace. Papermaking districts such as Yamato (Nara), Awa (Chiba), Tosa (Kōchi), Echizen (Fukui), Iwashiro (Fukushima), and Tamba (Kyoto and Hyōgo) were also soon noted for their *shifu*. Eventually it was made wherever paper was made and by all classes of people.

Farmers made a crude type of *shifu* in which both warp and weft were spun paper. This *morōjifu* often formed the underwear of the lower classes. The more refined *momen-*

*jifu* was *shifu* with cotton warp and spun paper weft, traditionally woven in one color such as indigo or with one colored stripe. *Momenjifu* was used in making covers for summer pillows and mattresses, coats, and mosquito netting. Also it was stencil dyed to make lovely obis and kimonos. The most refined *shifu* of all was *kinujifu* with a silk warp and spun paper weft. This expensive and precious variety, quite difficult to make well, was given in long rolls as gifts by the wealthy. *Kinujifu* is said to be deliciously cool to wear in summer.

*Shifu* experienced a major revival during World War II, when goods such as cotton cloth were scarce and people were required to cultivate great resourcefulness. Many army uniforms of *shifu* still exist in superb condition. After World War I the making of *shifu* had all but died out completely until around 1938–39, when Nobumitsu Katakura, a direct descendant of Lord Katakura Kojūrō, headed a group that conscientiously revived this ancient art in its old pure form once again in Shiroishi—and to this day is carried on there by himself and his daughter with paper supplied by Tadao Endō. The making of *shifu* has also been revived in Tamba by Michiko Kawaguchi, weaver and wife of a Buddhist priest.

### *Architecture—Shōji, Fusuma, and Byōbu*

IN ANCIENT times, the Japanese house was constructed as a single, large, open space surrounded by four outer walls. A residence of the aristocracy from the eighth century on was an open and fluid structure that permitted free circulation of air—and also dust, noise, people, insects, and pets, if any. As a result, many types of partitions to retain heat, stop drafts, and provide privacy were developed. Most of these were portable, some translucent, all relatively lightweight. Paper was an ideal material for partitions because of its lightness, heat retentive properties, ability to "breathe," to diffuse and reflect light, and to provide a surface for artistry. From this early use in partitions, paper has had a profound influence on Japanese architecture and thus on the lives and aesthetic concerns of the people.

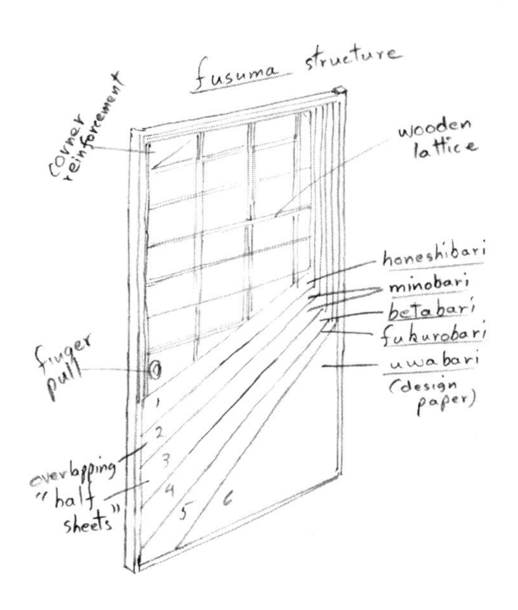


*Shōji*

Single-panel partitions mounted on a stand and utilizing paper were known as *shōji*. There were many types—*fusuma-shōji*, *karakami-shoji*, *akari-shōji*, and so on—depending on their use and construction. In time, *shōji* were inserted and fixed under lintels, and, later, grooved channels were contrived so the *shōji* could slide and be opened and closed. The *akari-shōji* ("illumination" *shōji*) developed into the form known today simply as *shōji*—thin, bright paper pasted onto one side of a light, wooden lattice frame. Used as windows or as room partitions, *shōji* have had a profound effect on the world's architecture.

*Fusuma*

*Fusuma-shōji* were single-panel partitions surfaced on both sides with thick paper or cloth and decorated with painting or calligraphy. *Karakami-shōji* were surfaced with "Chinese paper" printed with patterns in color and mica. These forms also developed into sliding panels running in grooves, and both types in time came to be known simply as *fusuma*. Today *fusuma* are used as sliding doors between rooms and as decorative sliding panels to cover storage alcoves.

The structure of a *fusuma* panel is a close cousin to or identical with the structure of a single leaf of a folding screen (*byōbu*). The aim of providing a light, portable, paper-faced panel that is strong, will not warp or wrinkle, and can be easily repaired or resurfaced may sound simple, but the making of *fusuma* and *byōbu* is a highly exacting craft involving great skill. A constant balance of tension must be achieved to keep the surface taut without the panel buckling.

The *fusuma* skeleton is a lattice of light wood in a frame of sturdier slats, reinforced at the corners and at the fingerpulls. Layers of paper are pasted over this; a common number is six layers, though this may vary. The first layer is a single large sheet, which is pasted onto the wooden frame and skeleton. The second and third layers are "half-size" (*hanshi*) sheets pasted at the top only and overlapping the sheet below. The fourth, fifth, and sixth layers again use single, large sheets the size of the panel. In the fourth, the entire paper surface is pasted and applied; only the edges of the fifth sheet are coated with paste and then the sheet is attached; for the sixth layer, the entire sheet is given a paste coat. Finally, decorative borders are attached to the *fusuma*. Different types and strengths of paper may be used for the various layers, but this depends upon the individual craftsman's preference. The balance of tension achieved between paper and wood is designed to give the panel strength and a long life, but if even a small member of the internal latticework weakens or breaks, the entire structure will eventually warp and must be taken apart and repaired. Like any craft, there are precision and sloppy craftsmen; it is an economy to employ the best skill in ordering *fusuma* and *byōbu*.

*Byōbu*

Screens of two or more (even-numbered) panels, six panels being the most common style, without a supporting stand, are known as *byōbu*. The history of the screen is old; there is a record of Emperor Temmu (reigned 673–686) receiving *byōbu* from the Silla kingdom of Korea. Decorative *byōbu* were a standard feature in the residences and palaces of the Heian aristocracy. Originally, each panel was decorated as a unit, but with later changes in architectural style and with the improvement in the method of hinging, the entire screen was used as a single, unbroken surface. These developments paved the way for the glory of the gold-backed, painted *byōbu* of the Momoyama period (1573–1615).

The uses of paper in Japanese architecture have had a profound influence on Japan, and, primarily in this century, on the architecture and interior design of many countries throughout the world. *Shōji* and *fusuma* are, in their stark clarity of line and austerity, timelessly modern and adaptable.

A people living in houses divided by sliding paper partitions and windowed with translucent paper screens will behave and think quite differently from a people surrounded by stone walls, stout wooden doors, and glass windows. The residents of a traditional Japanese house had little sense of separation from either nature or each other. In such a physical setting, privacy is largely an internal state; formality, deference, and quietude allow the smooth flow of everyday life. These qualities are an integral part of Japanese society today. In a small, interior space, walled with warm materials, the Japanese can relax—a person expands; the walls of such a room become like the skin of one's body. Big rooms and hard materials have the opposite effect—the person shrinks; he becomes divorced from his living environment. A masonry wall, however beautiful and secure, constantly states its immutability both as an enclosure and in its longevity. The very ephemerality of paper used in architecture allows the Japanese to relate to their

homes intimately in every sense. One knows the *fusuma* will wear out in a few years. The annual task of changing the *shōji* paper was a family activity. And, in warm weather, *shōji* and *fusuma* can be quickly removed from their channels to open the house to the outside.

The light coming through the *shōji*—the soft, diffuse light that hints rather than states—influenced house interiors and, further, the kind of objects that fit into these interiors. Potter Shōji Hamada tells of making a raku ware plaque with Japanese lead glazes he brought with him when he first went to England with Bernard Leach in 1920. But, he said, "It lacked something. The soft raku glaze colours blend well in a Japanese room of wood, paper, grass, and clay, but did not harmonize with a room of stone walls and glass windows; the surroundings were too hard." He proceeded to make a paper window screen, and, in its soft light, the effect of the raku panel changed. "Without such light and the appropriate materials surrounding it, raku ware was not satisfactory in England."[7] Hamada and Leach then turned to local materials and made slipware with English lead (galena) glazes; because of the nature of the local glazes and the way they were used, such slipware successfully fit into the English homes. It is not an exaggeration to state that the light from the paper *shōji* must have influenced the very essence of the Japanese aesthetic sensitivity and, in turn, the techniques of the craftsmen who made objects to use in Japanese rooms.

*Lanterns and Lamps*

Flexibility, strength, body, lightness, and soft translucency are qualities that have made washi the ideal substance for lanterns. When and how paper was first given this use in Japan has not been ascertained. The first paper shades were simple cones mounted around the heads of the standing oil lamps of the aristocracy and clergy. The paper cones diffused the light and protected the taper's flame from the wind. Yet the ribbed paper shade took hundreds of years to develop from this simple beginning, a fact surprising in a country in which ribbed fans and umbrellas were long and well known.

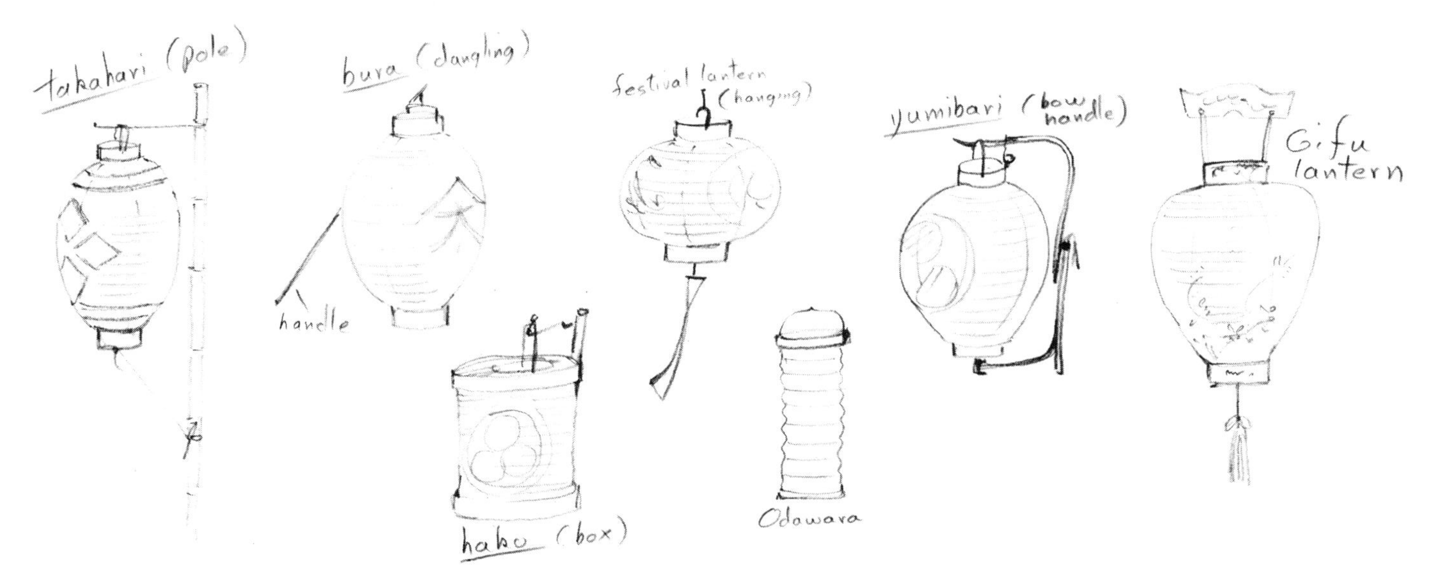


According to some sources, the folding paper lantern (*chōchin*) is said to have been introduced from China in the Muromachi period (1333–1573), seemingly by Zen priests. Other sources, however, suggest that collapsible *chōchin* were not made until the latter part of the sixteenth century, a time when *chōchin* styles were improved in general. By the Edo period they were in common use.

Special *kōzo* paper was made for *chōchin*, and centers of lantern production such as Gifu and Odawara grew up in proximity to quality paper-producing areas. The construction of the *chōchin* allowed a variety of forms and sizes, and numerous types developed, based on the basic spiral or hoop form of the supporting ribs.

Even with its many practical uses, the lantern did not lose its religious association. Lanterns of various types still play an important role in Japanese festivals, especially the Obon (All Souls) observances in summer and other summer festivals in which light is a central element.

Probably the most famous "lantern" festival is the Nebuta Festival of Aomori Prefecture, in which immense paper lantern-floats are constructed and painted into the likenesses of fantastic warriors, gods and goddesses, and monsters. In neighboring Akita Prefecture, towering arrangements of *chōchin* fixed about a central pole are carried through the streets. The men compete in balancing these poles on chin, forehead, shoulder, and back. Every town and hamlet in the Noto Peninsula has its *kiriko* festival, featuring vertical, rectangular "lanterns" borne on two horizontal poles. A *kiriko* may be so large that it requires thirty strong men to carry it or it may be just the right size for two toddlers to heft.

Contemporary sculptor Isamu Noguchi recognized the design possibilities of the *chōchin* in the early 1950s, when he began designing and making his *akari* lanterns. *Akari* means "illumination," and its character combines radicals for sun and moon. Made of quality Mino paper and utilizing the traditional lantern-making skills of Gifu, where a spiral rather than a hoop skeleton is employed, Noguchi's *akari* challenge the density of material itself and excite qualities of light. "The ideal of *akari*," he writes, "is exemplified with lightness (as essence) and light (for awareness). The quality is poet-

ic, ephemeral, and tentative. Looking more fragile than they are, *akari* seem to float, casting their light in passing."[8] With Noguchi's exceptional feel for materials, this is what they do.

Noguchi's designs are so strikingly simple and valid that they have enjoyed a constant popularity throughout the world, providing thereby an inexpensive and attractive lighting fixture. They go anywhere successfully, finding their way into rooms ranging from the humblest student lodging to the most sophisticated homes and apartments.

*Fans*

WHY the fan assumed such an important role in Japanese culture is still something of an enigma. Even in a country of hot, sultry summers, they do more than just push air—they are the essential sidearms of etiquette, the symbol of felicity and good fortune, respect, and all the refinements of Japanese civilization. As a result, the fan motif is ubiquitous, cropping up everywhere from textile patterns to gold lacquer decoration, from metal door pulls to ceramic designs, and even finding reflection in the shapes of such objects as food vessels.

There are two kinds of fans in Japan—the folding fan (*ōgi* or *sensu*) and the round, nonfolding type (*uchiwa*). Both have long histories. The *uchiwa* is the older of the two and is thought to be a Chinese import. The *ōgi* is believed to be a Japanese innovation, and this fan, in particular, reached heights of sophistication and artistry not found elsewhere in Asia.

The *uchiwa* has a stemlike handle and is paddle shaped. This type of fan exists throughout the world, and some say its shape developed naturally out of necessity in many warm climates. When it first appeared in Japan is not known, but a sophisticated form most likely was brought from Korea in the sixth or seventh century, if not before. During the Heian period (782–1184) the Chinese "drum style" *uchiwa*—a hollow frame covered on both sides with silk or paper—was known. The ribbed *uchiwa*, thought to be of Korean origin, became common in Japan roughly around the fourteenth century. The prosperity and ease of travel in the Edo period (1615–1868) saw the development of a great variety of regional styles of *uchiwa*, many of which are still made today.

The ribbed *uchiwa* is simply constructed—ribs radiate out from the top of the

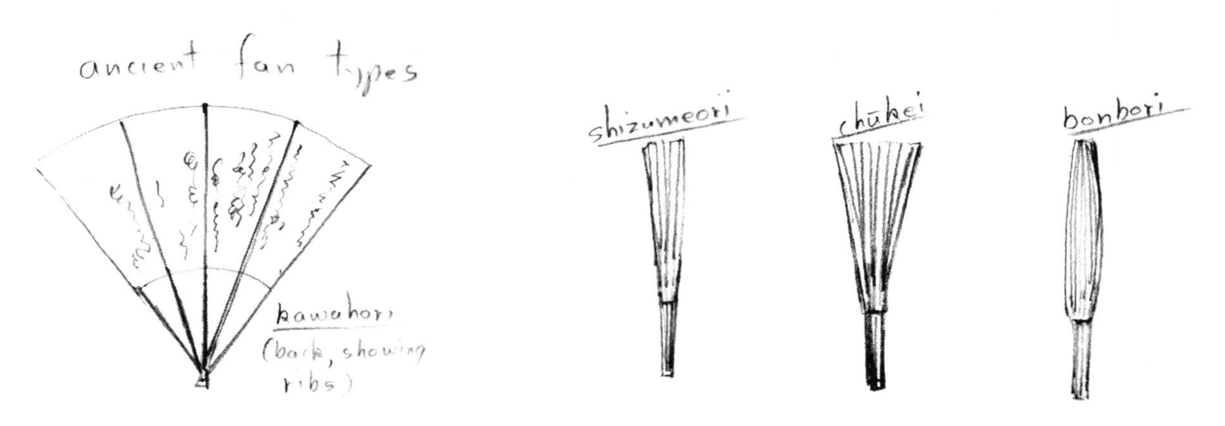


handle and are covered with paper. This form commonly utilizes a single stalk of bamboo, the appropriate length above a node being split to form the ribs; the stalk below this node is cut the proper length and shape to form a comfortable handle. The simplicity of its shape suggests it was inspired by a large, tropical leaf, and, in fact, certain types of palm fronds, dried and cut to shape, resemble the *uchiwa* exactly and are used for the same purpose.

Since *uchiwa* tend to be large and do not fold, they are found mainly in the home. Besides keeping one cool, they were used to fan kitchen and bath fires; large ones were employed to blow chaff from rice. *Uchiwa* are often brightly lacquered or coated with persimmon tannin (*shibu*) to waterproof them and lengthen their life. The practice of dipping a lacquered *uchiwa* into water on sweltering days may not provide a cooler breeze, but the effect is pleasing and refreshing. "Edo *uchiwa*" were decorated with *ukiyo-e* prints.

There are two major types of *ōgi*: the *hi-ōgi*, constructed of thread-tied cedar or cypress slats, and, developing from that, the papered fan, or *kawahori*. In the Heian period, the former was known as the "winter fan," and the latter as the "summer fan." An eighth century prototype of the former has been found, so its origins presumably go back to the Nara period, when courtiers first began using fans as an important wardrobe accessory. The papered and folding *kawahori* was developed during the Heian period, and by the end of the tenth century was being exported to China. It reached the peak of its development in the eleventh and twelfth centuries, when fans were essential accessories of cultural refinement and elegance. Both early types of *ōgi* were used primarily by the aristocracy. They were highly decorated, and many styles developed, some being associated with the various levels of court society.

The original *kawahori* was papered on one side only, and the five thin ribs could be seen from the back. The Chinese improved on this by inserting the ribs into the fan paper, making them invisible and providing two unbroken surfaces to receive decoration. This Chinese *ōgi* was imported into Japan and was again improved upon. The term *sensu* first appeared, along with various terms for different fan types, probably in the fourteenth century and today has become the word used for the common fan. *Ōgi* now has the implication of a folding fan used on formal occasions, notably ceremonies, in the Nō drama, and Japanese dance. Styles proliferated until the Edo period, when *ōgi* production was controlled by the shogunate and the forms became fixed.

In the meantime, the *hi-ōgi* grew in size and degenerated into a largely ornamental object. Its use largely died out in the Muromachi period, but today there has been a modest revival of fans delicately carved from aromatic woods such as camphor and cedar.

Chinese *ōgi* found their way to Europe, where the folding fan later became a fashion. Nineteenth century Japan-made export fans were modeled on the style produced in Europe. After World War I, surplus stock of this export style was sold on the domestic market; it shortly became the standard summer fan in Japan and remains so today.

Fan-making is a highly developed craft. The paper used must be strong enough to endure being constantly folded, snapped open and closed, and the constant battering that a fan is subject to. It takes years of experience for a fan-maker to acquire a knowledge of papers and glues and to learn the shaping and tapering of ribs that give the form balance, delicacy, and grace. Thin, fragile-looking papers of great strength are sought for fans; thus, *gampi* papers, even though expensive, are preferred fan papers. Historically, the amount of artistic ingenuity that has gone into the decoration of the fan's paper surface is overwhelming—whether painting, calligraphy, printing, or decorative embellishment of the paper itself.

*Umbrellas (kasa)*

THERE are two Chinese characters used for *kasa*: the first means a flared covering held aloft on a stick; the second means a covering worn directly on the head, that is, a flared hat. The first means, in effect, "umbrella," but is also applied to such things as mushroom caps and lampshades.

The introduction of the umbrella to Japan is still obscure. A prototype of the common umbrella (*bangasa*) was possibly a sixth century Korean import, but originating in China or perhaps in an area further west. Around 800 A.D., a kind of parasol called a *higasa* appeared, constructed of paper stretched over bamboo ribs, with a diameter of four to five feet. *Higasa* were carried by attendants to protect members of the aristocracy and high clergy. For centuries to follow, the Japanese, as the Chinese before them, used the umbrella, especially large red ones, in ceremonies and processions. Eventually priests of lower rank began carrying their own *kasa*; during the eighteenth century, everyone adopted the custom.

Quality washi is of prime concern in the making of *kasa*. Paper such as *Yamanouchi-shi*, *Hosokawa-shi*, and Kurodani's *shibu-gami* are strong, excellent umbrella papers that can withstand bending and handling. Washi is cut into isosceles triangles as long as the umbrella body. For additional strength, small, rectangular pieces are cut and glued where the bamboo ribs and spokes meet, then the triangular strips of washi are glued along the length of the rib spokes. In some umbrellas, wide paper strips cover several ribs each, but in the stronger, more finely constructed ones there is a strip of paper for each gap between spokes. *Shibu*, the processed juice of unripe persimmons, gives the paper strength and makes it water resistant. The *shibu* may be applied by the paper-maker or by the umbrella-maker. Paper umbrellas for gardens and for carrying in snow and rain are coated with a high-quality drying oil (*kiri* or *egoma* oil). This adds flexibility to the paper and increases its water-repellent quality.

*Kites*

CHINA is assumed to be the home of the kite, for records of kite flying there date to the Han dynasty (206 B.C.–220 A.D.). Originally, Chinese kites were used for surveying and other practical purposes, but eventually kite flying also came to be enjoyed for pleasure and as a children's pastime. Japan learned of this Chinese flying device during the Heian period (794–1184), it is thought, possibly from the Koreans. Paintings and legends of kites suggest that, quite early in their history in Japan, kites were associated with temples.

The most fantastic part of kite history is the kite's use as a human carrier. Records relate how enormous kites were medieval military carriers, bearing spies into and out of cities and camps under siege. They were also used as observation posts—man's first reconnaissance flights.

In the Edo period, the kite became a children's amusement, customarily flown during the bright days of the New Year's celebration. Yet, the kite retained its place in the adult world too. The Japanese sought to make larger, more majestic kites, which, by the eighteenth and ninteenth centuries, had become so large that kite devotees needed the length and strength of ship tow ropes to control their aeronautical constructions. The *wan-wan* kite of Shikoku appeared at the end of the last century with an innovative structure employing a particularly tough *kōzo* paper. One *wan-wan* made in 1906 was 20 meters wide, had a tail of 146 meters, and weighed two tons. It required 150 men to launch and fly.

Today, in various parts of the country, districts within a town or small neighboring towns compete in battles of giant kites. The tails of these kites are studded with small knives and pieces of broken glass as well as whistles and noisemakers. Each team launches its kite on opposite banks of a river and then tries to cut loose the other team's kite.

Japanese kites have been and still are made of washi, for it is strong, flexible, porous, and lightweight. Kite paper is nearly always a variety of *kōzo-gami*, sometimes cut with a portion of chemical pulp. Some have said that pure *kōzo* kite papers are stronger wet or damp than dry, but I have no experience to substantiate this, and rather doubt it. However, kite-makers, especially those producing the giant kites for kite battles, are most particular about the paper they use. They are naturally reluctant to reveal paper types, since this information is a trade secret. Traditionally, kite papers were most frequently selected from the better quality, pure *kōzo* papers native to the maker's area. Such paper must be able to withstand enormous stress in wind, sometimes rain, and general knocking about; therefore medium-thick, pure *kōzo* papers such as *Nishinouchi* from Ibaragi Prefecture and *kōzo* papers from Iiyama, Nagano Prefecture, are excellent. Umbrella papers are often used as well, although for kites they are not oiled or waterproofed. All kites in Japan are painted—plain white being symbolic of mourning—and since some of the designs are outlined in hot wax, this in effect moisture-proofs the paper somewhat.

The paper is pasted directly onto the bamboo splints forming the backside of the kite's structure. Wheat flour paste, and sometimes rice paste, are favored. A few kitemakers wrap the bamboo bones first in thin *kōzo* paper, then paste the kite paper onto the wrapped skeleton; this makes a stronger kite, since paper is pasted to paper rather than to bamboo.

The *koi-nobori*, or "climbing carp" streamers, are favorite decorations in Japan and abroad. Not really kites but wind socks, these streamers depict the indomitable carp who fight their way upstream to spawn, thus symbolizing courage and manliness. Once made only of paper, cotton carp streamers were introduced in the late nineteenth century; today they are largely of synthetic fabrics.

### *Dolls and Toys*

PAPER was one of the major materials used in making toys and dolls. Often recycled sheets and pulp were formed and painted, but the brightly printed *chiyogami* of Edo and Kyoto and also the crisp strength of newly laid *kōzo* paper were fully utilized. The paper toys of Japan are a world unto themselves, and the making of traditional, local paper toys has recently seen a great revival. Many urban housewives and

girls with the spare time for a hobby have taken up the making of sophisticated paper dolls and figurines—both traditional and modern—and provide a steady and major market for the makers of highly decorative colored papers. But perhaps no papier-maché figure is as ubiquitous and well loved as the Daruma.

The Daruma figure—both the toy and the good luck doll—is found everywhere in Japan, a caricature of red-robed Bodhidharma, who brought Zen (Ch'an) Buddhism from India to China. The son of a king of southern India, this Buddhist monk traveled to China sometime between 470 and 520 A.D. According to legend, he spent nine years alone in meditation facing a wall. So long did he sit, it is said, that his legs withered away. But he achieved enlightenment. The usual form of the Daruma toy is weighted at the bottom and always rights itself when pushed over. Like Bodhidharma, it represents dauntless perseverance and persistence.

This doll came to Japan sometime in the Muromachi period (1333–1573). Although wooden Daruma are common, the papier-maché dolls are the most popular. Layers of pulp or sheet paper and glue are applied, dried, and built up over a clay or wooden mold. The papier-maché form is then cut in half, removed from the mold, a weight inserted (if desired), the halves joined together again with paper, and the form painted. Recycled paper is usually used.

In some towns in Japan, papermakers devote themselves to the manufacture of papier-maché sheeting, which consists of various mixtures of recycled machine and handmade papers, pure fiber *kōzo* and/or *mitsumata*, and chemical pulp. The gray sheets are dried on the grass for softness. It is amazing that such paper is still made by hand, with care and pride, as I have seen in Yamanashi and Saitama prefectures.

Different areas of the country or even towns have their own styles of Daruma. Gumma Prefecture and the Sendai area produce the most commonly seen, the eyeless Daruma of good fortune. When purchased, a wish is made and one black eye painted in the belief that a one-eyed Daruma will intercede for the owner in order to get another eye. When the wish is fulfilled or success comes, the other eye is painted in. This Daruma is a favorite good luck charm, used particularly by politicians during elections and by new business owners.

At the end of every year in preparation for the New Year, which is considered an auspicious time for wishing, a Daruma market-fair is held north of Tokyo. This tradi-

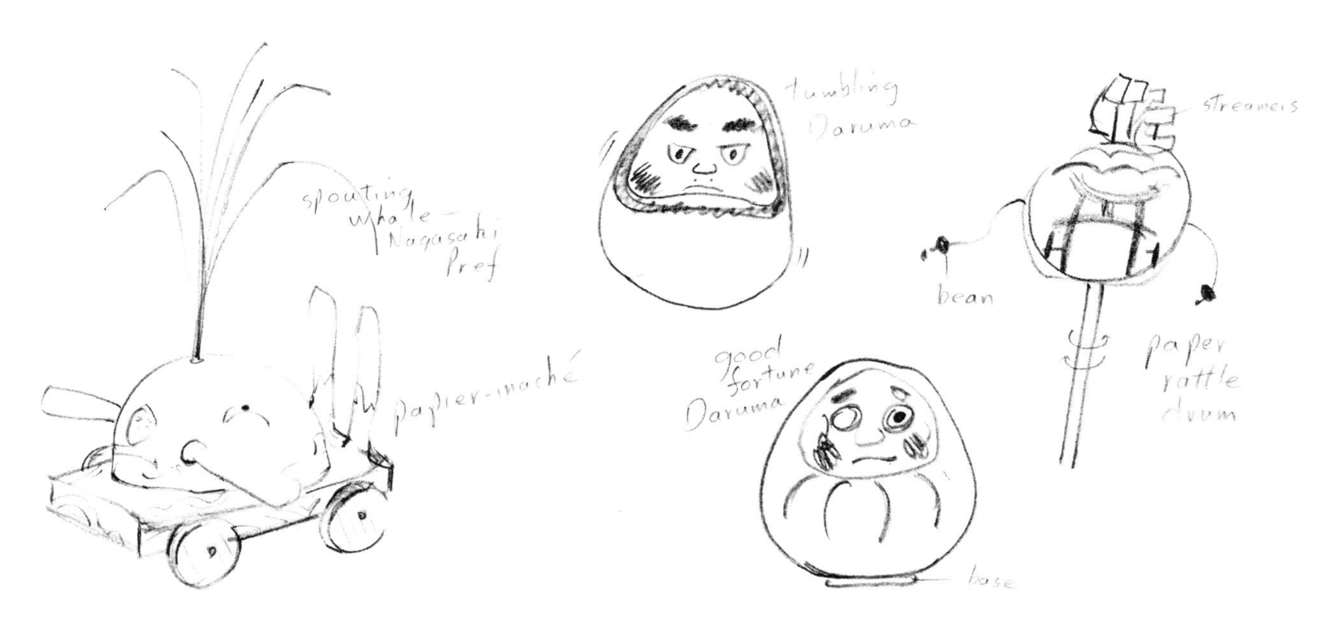


tional market, carried on for several centuries, attracts people from near and far, who come to sell and buy Daruma from thumbnail size to some so large they must be carted home by pickup truck.

*Koyori and Mizuhiki*

THE origins of *koyori*, twisted paper string, and *mizuhiki*, the more elaborate paper cord moistened and twisted on a frame, are not well known, although scholars agree that the processes behind them are quite old. It is probable that they developed about the same time as, or just prior to, the invention of *shifu* around 1600, since the ideas and techniques involved are essentially the same.

*Koyori* string is made from a narrow sheet of paper held by one corner and rolled on a flat surface or twisted by hand to form a strong cord. Or, a paper sheet may be cut into one long continuous strip and spun in the same manner as *shifu* is made. People wove *koyori* into thousands of useful shapes. By lightly pasting the cord over a mold, or by weaving it into basketwork shapes and heavily lacquering them, the Japanese created an innumerable variety of useful objects—bags, gunpowder and water flasks, hats, baskets, tobacco pouches, saké cups, purses, boxes, and even more ingeniously created items. The lacquered *koyori* wares are both inflexible and waterproof. Many old examples can be seen today at the Ōji Paper Museum in Tokyo, and some can still be obtained in antique shops and markets.

*Mizuhiki*, although quite similar to *koyori*, is finer cord manufactured by a rather more elaborate technique. A narrow sheet of paper is twisted diagonally from one corner, plunged into water quickly, twisted more and stretched either on a frame or a wheel; or it is fastened to a vertical pole and twisted and stretched horizontally outdoors, rather like a clothes line. There it dries tightly in the sun. *Mizuhiki* were for centuries used as elaborately twisted hair ties for both sexes until shorter hair styles became popular. Today *mizuhiki* are commonly used for wrapping and decorating gifts, with very intricate rules attached to the etiquette of their colors and the methods of tying them.

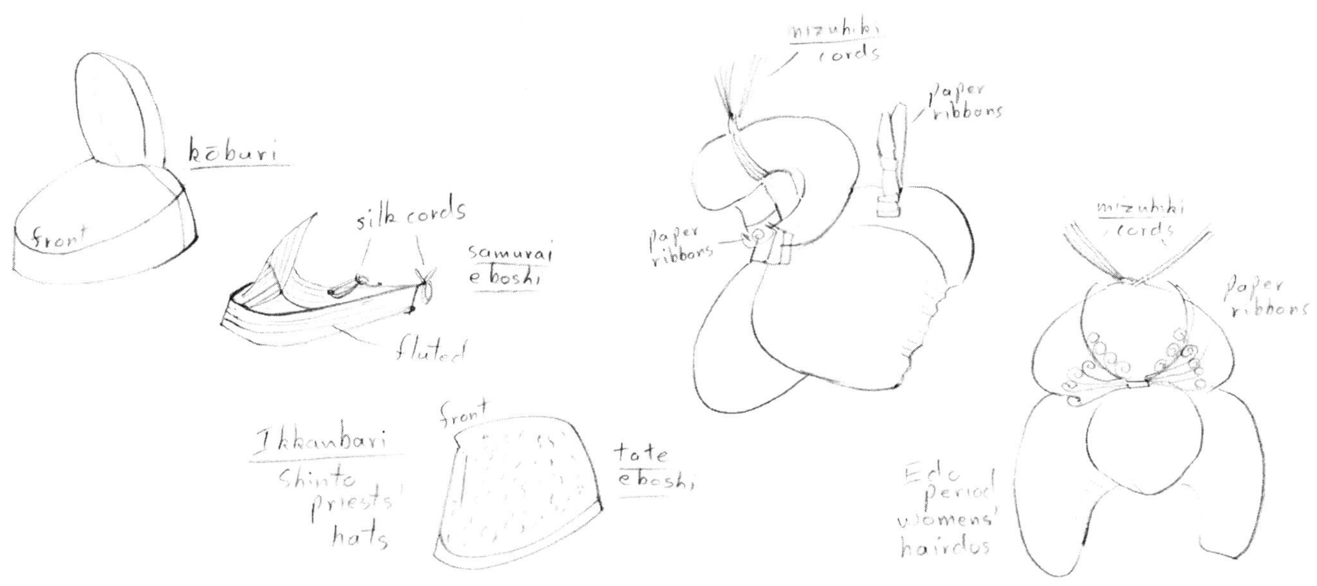


### *Lacquered Papier-Maché (Ikkanbari)*

LACQUERED objects formed from twisted and woven paper string (*koyori*) have been made in Japan for centuries. Lacquered papier-maché items may have a longer history, but are considerably rarer. Usually, layers of sheet paper were applied over a mold, and the finished form was then lacquered. Most frequently made were boxes, trays, and containers of various kinds. Pure fiber pulp was used to form such objects as well. The term *Ikkanbari* comes from the name of Hirai Ikkan (1578–1657), a Chinese lacquer craftsman who immigrated to Japan and was patronized by the Tea master Sen Sōtan. It is probable that lacquered paper objects were made in Japan before this time, but little research has been done in this field yet. Paper objects—lacquered, oiled, and plain—were made from early times in great variety (including furniture) in Korea. A scrutiny of the history of paper objects of both Japan and Korea might yield some important insights into this neglected craft. In the late nineteenth century and early decades of the twentieth, kitsch and cheap objects were also made of lacquered paper, both for export and domestic consumption.

NOTES

half-title: Tokutarō Hanada and Kiyofusa Narita, *Kami-suki uta,* 2 vols. (Tokyo: Seishi Kinenkan, 1951). (group trans.)

1. Until 1966, the Japanese *dhāraṇī* were considered the world's first printed texts. However, a Korean *dhāraṇī* sealed into a stupa in 751 was found at Pulguk-sa temple in Kyŏngju, and this is now the oldest known example of printing.
2. Seikichirō Gotō, *Washi no furosato* (Tokyo: Bijutsu Shuppansha, 1967). (trans. by David Hughes)
3. Bunshō Jugaku, *Paper-Making by Hand in Japan* (Tokyo: Meiji Shobō, 1959).
4. Shin'ichi Kannō, *Shiroishi-gami* (Tokyo: Bijutsu Shuppansha, 1966).
5. Munemichi Yanagi, "Creative Art from Paper," *This is Japan* 18.
6. Mock Joya, *Things Japanese* (Tokyo: Tokyo News Service, 1968), p. 656.
7. Bernard Leach, *Hamada, Potter* (Tokyo: Kodansha International, 1975), pp. 51–52.
8. Isamu Noguchi, *A Sculptor's World* (New York: Harper and Row, 1968).

hariko dog

*Good paper is made in quietness—*
*I confront the paper mold, composed.*

*Keeping my mind like fresh water,*
*I can make pure paper.*

# How Washi is Made

Traditionally papermaking has been a family or community concern, carried on in winter by farming people between the harvesting and the planting. Because farms have always been small in Japan, as was the farmer's income, making paper proved early in its history to be a good sideline during the winter's comparatively leisure months—especially since fine papermaking requires cold weather.

After the rice has been harvested and persimmons hang ruddy and ripe on leafless trees, the papermakers begin cutting down the shoots that supply them with fiber. Before the November or December frost, they are scooping the pulp solution on the bamboo screens, and by the time the first snow falls, drying boards covered with crisp, new winter paper reflect the sun across the hills. While papermaking used to be only a winter activity, the discovery of preservatives retarding spoilage of raw materials has made the work possible year round. Today many towns across the country have become full-time papermaking communities.

Japanese soil is rich; water and sun abound, and the humidity is high. But most of the land is mountainous—only twenty percent of it is arable. Consequently both farmers and papermakers alike have always had to put what land there was to optimum use in raising their crops. The farmer has ingeniously taken the rolling land and terraced it into rice fields. The papermaker has searched for wild varieties of paper fiber trees; where these plants were scarce, he has grown them in the lowlands, or, like the farmer, he has terraced the hillsides for their cultivation.

Papermaking was a craft suited to the climate and lay of the land. It especially thrived in mountainous areas where conditions were favorable, where wild bast fiber shrubs grew plentifully. Here the best traditions of paper technique arose and can still be found almost untouched by modern life. Their product has always been in demand.

*Water and Cold*

Aside from the growth of the paper fiber trees, Japanese papermaking is dependent upon the land's generous bestowal of two essentials—water and cold. Water is the necessary element in every phase of papermaking. To be at its best, papermaking needs water that is pure and cold and moving; this is why it is said that mountainous areas produce the very best paper. Japan's rugged terrain is cut by numerous rivers, above and below ground, and fresh mountain streams fed by melting snows, all excellent sources for clear, cold waters free of bacteria, minerals, and impurities. Soft water gives paper a lucidity that hard water destroys. Minerals in the water, especially iron and manganese, will spoil the complexion of paper over a period of time.

Cold is the second essential. The material used for paper is, figuratively, a living substance even after it is cut; its enemies are whatever promote its death or decay—largely bacteria in impure water coupled with a hot, humid atmosphere so characteristic of the Japanese summer. These breed bacteria on the damp fiber. In the cold Japanese winter, low temperatures maintained in the unheated workshops help retard that bacterial growth.

Coldness also tightens the plant fibers, forcing them to contract, and so produces a paper that is fresh in appearance, crisp, and strong. In contrast, summer paper tends to be limp and fairly weak. In northern areas where bleaching of the bark is accomplished by such primitive methods as immersion in a cold, running stream or half burying it in the snow, the finished paper is exceptionally chaste and beautiful. Cold also works in the paper solution vat to promote the effect of the vegetable mucilage.

### *Climate and Nature*

MOST of Japan's three major islands where paper is made, Honshu, Shikoku, and Kyushu, lie between the thirty-fourth and forty-first parallels. Thus the country's climate is generally described as temperate, but there are important differences. For one, located between Siberia to the west and the great Pacific to the east, in winter the islands are victim to capricious and frigid winds bearing crushing snows from the northern plains of the continent, at the same time that the southern and eastern shores are washed with the warm and fertile Black Current flowing north from the equator. Like India and Southeast Asia, Japan receives monsoon rains. These the Japanese call the "plum rains"; they arrive in mid-June and stop abruptly toward the end of July. In their wake the country experiences a summer and early autumn of intense heat and suffocating humidity. Good paper is rarely made at this time—both the materials and the workers lose their vitality. In autumn come typhoons, with gale winds and rains that tear away at forest, field, and mountainside. And so while Japan may have a temperate climate, it is dramatic, incorporating stark contrasts and elements of both the tropical and the frigid.

The drama of weather and climate in Japan has in some ways left the Japanese a receptive and resigned people. Heat and humidity are a devastating combination, for the resistance needed to overcome them is formidable. But the violent and unpredictable force of typhoons, tidal waves, earthquakes, and volcanic eruptions have tempered the "passivity" of the Japanese to a certain determined resiliency and a sometimes combatant spirit, much like bamboo growing in a northern climate, bowed but not broken with the weight of snow—soon it snaps back. Perhaps this is one clue to the Japanese dedication to work, for nature is Japan's desperation as well as the fountainhead of abundance and fertility. The people know their survival lies in the ability to anticipate nature's quirks and, rather than attempting to tame nature, to understand deeply her every aspect.

The rhythm of the growth of crops always directly affects the rhythm of farmers' lives. In Japan that rhythm could most characteristically be described as *urgent.* The seasons change with startling rapidity, and it is vital that the farmer keep pace with them. With the same sense of urgency that compels a farming family into a sudden burst of effort necessary to harvest their rice crop in ten short days before the typhoons hit, the papermaker traditionally was under extreme pressure to beat the boiled fibers and scoop them into paper all in a few days before they would rot, particularly in the heat and humidity of a Japanese summer.

It was the exception rather than the rule in the days before World War II for a whole family to complete a full day's labor processing the fiber, as well as their farming, then sit up all night without sleep, beating the fiber ceaselessly across a low table; and then without rest or interval, spend the next day scooping it into paper. Although papermaking today is usually not quite so austere and harsh, it is still arduous, sporadically urgent work. If it suddenly rains or snows on a day when the paper must be dried, precious time and labor is lost trying to dry the sheets indoors. Like the farmer, the papermaker's work is ever hampered by the fickleness of climate and fury of weather.

Fortunately this urgency is an acute condition rather than a chronic one—the weather is often mild in Japan, especially in winter on the Pacific side. There, winter is the dry season, sunny and mild, a time when nature takes its breathing space. In fact, during the day it is more invigorating and warmer to work outdoors in the sun than to work inside in the dimly lit and poorly heated rooms. So the papermaker's desire to work outdoors or in semishelter is a natural and a pragmatic one and has brought him closer to the earth than his Western counterparts.

*Materials*

THE first Chinese papers were made from tree bark, hemp, fish nets, and old rags in various combinations. It is not well known from what the first paper that came to Japan via Korean merchants and seamen was made, but it was probably a variety manufactured from hemp rags. The bast itself was evidently crudely processed, for the paper was brittle and usually short-lived; but it is noteworthy that such techniques as adding starch and gypsum to the vats and sizing the sheets with vegetable gelatins, then dyeing and polishing them, were already well developed when the paper first reached Japanese shores. After the Korean monk known as Donchō in Japan introduced the latest skills to the imperial court in 610 A.D., Japanese craftsmen set about experimenting with techniques and with their own native plants. From the start, the Japanese papermakers used the raw hemp fiber itself for the bast rather than hemp cloth rags, and so, inventing fresh skills that grew out of this new approach, developed a style of papermaking that always was to remain distinctly Japanese.

In Japan today, three fibers are used for almost all washi: *kōzo* (paper mulberry), *mitsumata*, and *gampi*, in order of their importance and use. All are small trees that once grew wild and plentifully throughout Japan. Now, after centuries of harvesting, the wild varieties have mostly disappeared from the countryside, but the farmers have replaced them with the superior domesticated varieties of *kōzo* and *mitsumata*. The *gampi* tree is vanishing.

Each of the fibers has its own personality and characteristics. Here is how Sōetsu Yanagi saw them:

> *Gampi*, *kōzo*, and *mitsumata* form the trinity of materials for Japanese paper. All the various kinds of washi are made from these fibers. *Gampi* commands the highest position of the three, with *kōzo* on its right and *mitsumata* on its left.
>
> No beauty can be compared to the beauty of *gampi* in quality, richness, and dignity. It lasts forever. Softness and hardness, truth and falsity—all things meet in it. There is no paper in the world as noble as this.
>
> *Kōzo* is the masculine element, the protector in the world of paper. Its fibers are thick and strong; it will endure the roughest handling. Because of *kōzo*, washi retains its vitality even today. If *kōzo* did not exist, how impotent would be the world of paper.
>
> By comparison, *mitsumata* is the feminine element, lending its softness and supple qualities. There is no paper as graceful as this—its nature is delicate, its

complexion soft, and its disposition modest. If there was no *mitsumata*, there would be less charm and feeling in the world of paper.

*Gampi*, *kōzo*, and *mitsumata*—these three all work together to protect and nourish the life of washi. Depending on one's need or taste, one may choose among the three. No matter which you select, you will find in each the distinct beauty of washi.[1]

*Kōzo*

THE term loosely and colloquially applied to a variety of papermaking mulberries in Japan is *kōzo*. Prince Shōtoku (regent 593–621 A.D.), among countless other reputed feats, is said to have introduced *kōzo* as an important papermaking fiber, although it is just as likely that this was the fiber used by the Korean priest Donchō when he first introduced papermaking techniques to the imperial court. Regardless, *kōzo* was one of the first fibers to be used successfully for paper in Japan after hemp, and it was with great foresight that Prince Shōtoku encouraged its widespread cultivation. In a short time, several varieties were tried, and many proved to be excellent for paper.

Before the introduction of cotton to Japan, *kōzo* fiber (as well as hemp fiber) had already long been in use for the weaving of cloth, so it was perhaps natural that its excellent bast properties would also make it well suited for paper.

*Kōzo* is the toughest, strongest, and roughest of all the Japanese paper fibers; it is the most distinctly masculine. The fibers are long and sinewy. They neither shrink nor expand, which makes *kōzo* paper excellent for printing. *Kōzo* paper has a somewhat uneven surface, rather porous and highly absorbent in its unsized state. Its character is malleable—a good papermaker can make it soft and fluffy, or resilient and crisp, or coarse-complexioned and thick. Its strength, natural warmth, and modest price make it the most versatile of the handmade papers.

Several varieties of *kōzo* grow throughout the country; therefore, until recently, simple *kōzo* paper varied considerably in character from region to region, and many of these regional differences are still retained. In the northern part of Japan, *kōzo* fiber is narrow and lacks stiffness and toughness, but such is the best fiber for *hanga* (woodblock print) and *shōji* (latticed window) paper. The farther south it grows, the rougher the fiber and the harder it is to make good paper from it. Today, due to good trucking and communication, many kinds of *kōzo* and other fibers are available to papermakers all over the country, along with the knowledge of how the various areas process them.

There is a considerable amount of confusion surrounding the term *kōzo*, for

usually it is applied to only one of several rather distinct genera of mulberry (family *Moraceae*). Moreover, besides the fact that each region of the country has its own vernacular terms for these plants, there is a general confusion over the various species, all of which papermakers loosely refer to as *kōzo*. I have divided the three most widely used species within their genus (*Broussonetia*), describing the growth and characteristics of each. The species that grows in the north is *Broussonetia kajinoki*, Sieb., casually known as *kōzo*. This is not to be confused with *Broussonetia papyrifera*, Vent., which generally goes by the colloquial term *kajinoki*, but sometimes by *kōzo* as well. The third type, *Broussonetia kaempferi*, Sieb., is called *tsurukōzo* as well as *kōzo*. Other terms include any of these mulberries—wild *kōzo* is called *yamakōzo*, literally "mountain" *kōzo*; the cultivated kind is called *nokōzo* or *satokōzo*, "field" or "village" *kōzo*.

*Broussonetia kajinoki*, Sieb.

This is considered the best of the various *kōzo* plants for making paper. It grows wild, but it is also cultivated all over Japan, particularly thriving in the cold, northern regions. This mulberry is thought to be indigenous to Japan, although in very ancient times it may have been introduced from Taiwan or other points south.

Like hemp and other mulberries, *Broussonetia kajinoki* is a member of the *Moraceae* family. This small tree grows to a height of from six to sixteen feet. Its leaves are largish, alternate, serrated, and deep lobed with pointed tips and silky hair on both sides. Male and female flowers grow on the same stalk, with their stamens and pistils on separate flowers. In early summer the tree bears a sweet, purple fruit. There are at least three to four varieties, differing in bark color and trunk markings.

This species of *kōzo* is rather short fibered compared to *Broussonetia papyrifera*. In ancient times the Japanese wove clothing from it. It is well suited for making such fine papers as *tengujō* (see p. 190).

*Broussonetia papyrifera*, Vent. (*kajinoki*)

This species is often confused with or regarded the same as other species of *kōzo*. Unlike *Broussonetia kajinoki*, this variety is not indigenous to Japan. It is a native of Burma, China, Thailand, and Polynesia. In the South Sea islands it has long been used for making tapa and kapa cloth, and in Java for a rough kind of paper formed into such

numerous things as quivers and bedding. The paper mulberry, as it is also called, is basically a tropical plant, although it grows throughout most of Japan and some parts of the U.S.

This tree grows to a height of about twenty-five to thirty feet, branching near the ground. The deciduous leaves are divided into three or five lobes when young and generally become an oval shape later on. Distinct male and female flowers grow on separate stalks. Female flowers are marblelike balls, while the male flowers are drooping, cylindrical catkins. After three years the paper mulberry bears very sweet, dark red berries.

Compared to *Broussonetia kajinoki*, this species is long fibered and tough.

*Broussonetia kaempferi*, Sieb. (*tsurukōzo*)
This fiber receives only brief mention, since it is now in little use and is considered an inferior mulberry fiber. *Tsurukōzo* grows wild and only in Kyushu. Its colloquial name means "winding" or "vine" *kōzo*.

### *Gampi*

THE family *Thymelaeaceae* incorporates the rest of the important papermaking genera—*Edgeworthia* (including *mitsumata*); *Daphne*, including some fibers used for papermaking in the Himalayan area; and *Diplomorpha*, numbering several species of *gampi*. There are some eight or nine species of *gampi*, all reputedly good for papermaking, but the one considered the best is *Diplomorpha sikokiana*, (Fr. et Sev.) Honda.

*Gampi* was one of the early fibers used for papermaking, its history stemming from the Nara period. *Gampi* fiber is extremely tough, and for this reason *gampi* paper is sometimes also known as *shifu*, or "paper cloth"—although not to be confused with true, woven *shifu* (see pp. 56 and 211).

*Gampi* fibers are long, thin, fine, and somewhat shiny. It is said to be the trickiest of the fibers from which to make paper. Its fibers are extremely strong. Paper made from *gampi* is smooth and lustrous. Like the *mitsumata* fiber, *gampi* contains a bitter chemical repugnant to paper-eating insects. It is not only nonabsorbent, but also damp-resistant. Experts say that *gampi* paper lasts forever.

*Diplomorpha sikokiana* grows to a height of about five feet. Its outer bark is dark brown and shiny. The *gampi* leaf is ovate, with both sides covered with silky hair. Small yellow flowers blossom in early summer. *Gampi* is never steamed before removing its bark—to do so would damage it. The plant is harvested in February through May, when it contains the most water; this is the time when the bark is easiest to peel off.

*Gampi* grows wild in mild climates, but stubbornly resists attempts to cultivate and domesticate it. The supply of wild *gampi* that used to grow so plentifully on Japan's hillsides is now nearly depleted. Some *gampi* bark is now being imported from Korea and China, but most papermakers consider it inferior to the Japanese stuff. Today making *gampi* paper is a specialty of increasingly fewer papermakers. It is the most expensive of all the Japanese papers, becoming a rarity of the paper world.

### *Mitsumata*

ALONG with *gampi*, *mitsumata* is also of the *Thymelaeaceae* family, genus *Edgeworthia. Edgeworthia papyrifera*, Sieb. et Zucc. is the plant known as *mitsumata*, with a large-leafed variety growing mainly in Shikoku, and the small-leafed one especially abundant in the Suruga area around Mt. Fuji. Neither variety flourishes very far north—usually below the thirty-eighth parallel. *Mitsumata* grows tall in areas of plentiful rainfall, but the inner bark is thinner; when grown in drier areas, the bark is thicker. Today it is particularly cultivated in western Japan.

*Mitsumata* was discovered rather late as a paper fiber. Scholar Bunshō Jugaku offers the theory that Shuzenji in the Izu Peninsula was one, if not the first, place to use it extensively in making paper, around the late 1500s. Seikichirō Gotō told me a local legend of the Mt. Fuji area concerning *mitsumata*. A Mr. Watanabe, great-grandfather of the present mayor of Fujinomiya, was climbing Mt. Fuji one day in the late 1700s. The rope that supported his backpack broke. Looking for something to use as a substitute, he found the *mitsumata* tree; its bark proved supple and strong. Later he thought this might be a good fiber for paper and thus introduced the idea to local papermakers. Now *mitsumata* is cultivated in many temperate areas of Japan.

*Mitsumata* fibers are soft and quite absorbent. Closely inspected, a piece of dried inner bark appears to have a spongelike network, found in none of the other papermaking barks. *Mitsumata* fiber is rather short compared to *kōzo* and *gampi* and therefore weaker than those. Not infrequently it is used in combination with one or the other; it is used with bamboo in the manufacture of *gasenshi* (see p. 175) or with wood pulp in a modern-day version of *torinoko* (see p. 191). Pure *mitsumata* paper is ever so slightly reddish or orangey in tone, fine-grained, soft, and with a slight luster. It is reputedly the easiest fiber with which to work. Like *gampi*, it resists paper-eating insects.

*Edgeworthia papyrifera* grows to a height of from three to ten feet. *Mitsu* means "three" in Japanese, and *mata*, "pronged" or "forked"—this describes the plant's most distinguishing characteristic, the stalks' dividing into three branches at every joint. *Mitsumata* leaves are light green, long, narrow, and undivided. In southern Japan the buds sprout in early winter and bloom into fluffy, acrid, cream-colored tufts of flowers in March before the leaves open. I have seen the shoots being steamed to loosen the bark in Yamanashi Prefecture in March—as the steamer shed door was opened to remove the steamed shoots, a thousand white blossoms tumbled out over the ground. The shoots are harvested twice in winter, new shoots growing and being cut every other year from the same plant for eight to twelve years. After that, they must be replaced by new plants grown from seed.

In general only second-grade *mitsumata* is available to papermakers. The highest quality bark is bought up by the Japanese government every year and used in the manufacture of bank notes.

### *Other Papermaking Fibers*

EXPERTS have remarked that paper can be made from almost any kind of vegetable matter—as long as cellulose is present, there is the possibility of paper. Early in the history of Japanese papermaking many plants were used for paper besides *gampi* and *kōzo*. We know from documents in the Shōsō-in, the eighth century Imperial Repository in Nara, that numerous fibers were tried and combined during the Nara period in search for the best raw materials for paper.

### *Hemp* (*asa or taima*); *Cannabis sativa*, L.

HEMP paper is the oldest kind of true paper in the world. The paper first introduced to Japan from China was manufactured largely from hemp rags. In Japan's Nara period, hemp was the most important papermaking fiber. Many of the celebrated printed Buddhist charms enshrined in one million miniature wooden pagodas in 770 A.D. are of hemp paper (see p. 42). Gradually its use declined at the end of the Nara period as the increasing demand for paper became more efficiently met by *kōzo* and *gampi*. Besides, its surface is rough compared to these two papers and therefore inferior for writing and printing. To a very limited extent *mashi*, hemp paper, is still made today, although the methods and materials used to make it are very different from the ancient *mashi*. Hemp paper is supple, lustrous, and strong. It is considered a high-class paper in spite of its defects and from ancient times has been symbolic of the sacred in Japan.

### *Wood Pulp*

WHEREAS all the major Japanese papermaking fibers come from deciduous trees, wood pulp is from conifers, usually imported from the U.S.S.R., U.S.A., and Canada. Wood pulp is the base material of machine-made paper, in Japan as everywhere else. The entire core of conifers such as spruce, pine, fir, and balsam are reduced by sulphite processes to a mostly cellulose content, then bleached and formed into stiff sheets used primarily in machine-manufactured paper. Wood pulp fibers are weak, short, and replete with chemical residues—naturally inferior to pure, processed Japanese bark fibers. Unfortunately some of Japan's contemporary hand papermakers "fill" the paper stock solution with wood pulp fiber in varying percentages because the pulp is cheap and convenient (although more expensive today than in recent years).

Some argue that wood pulp, when used in combination with other fibers, makes the paper soft, thick, and pliant, and thus is especially good for such as woodblock print papers. But the best papers contain no wood pulp, and the relative absence of chemical residues in the sheets greatly prolongs the paper's life.

### *Rice Straw* (*wara*)

ALTHOUGH "rice paper" is a Western misnomer of long standing, it is not entirely true that washi has nothing to do with rice, for rice straw has long been used in the manufacture of some inferior types of Japanese paper. Also, small amounts of rice powder loads were once added for luster and to fill the paper's pores. Straw in the old days was comparable to wood pulp in today's papermaking in its role as a cheap "filler." To a much lesser extent it is still used today. *Wara* is quite short-fibered, brittle, weak, and therefore generally an inferior material; however, due to its cheapness and absorbent powers, it is practical for *wara-banshi*, which is practice paper for calligraphy, and even an important ingredient in good quality *gasenshi*, paper for Chinese brush painting. *Wara* must be used in combination with other fibers, and then sparingly. It is too weak to stand on its own.

The straw is first processed by cutting the joints out of the stalks. The top of the stalk is the best part, the chaff the worst. Because straw contains such a low percentage of cellulose, a large amount of caustic soda is used in boiling it. After chemical digestion, it is pounded until pulpy, about one and one-half hours. Before the days of mechanical beaters, the *wara* was placed in a large stone mortar and a lever connected to the pestle was tread upon in a manner similar to the earliest Chinese papermaking techniques.

In the town of Misumi, Shimane Prefecture, Kiyokatsu Shimizu combines cotton, bamboo, *mitsumata*, recycled paper, and straw to make a very fine *gasenshi*.

*Bamboo* (*take*)

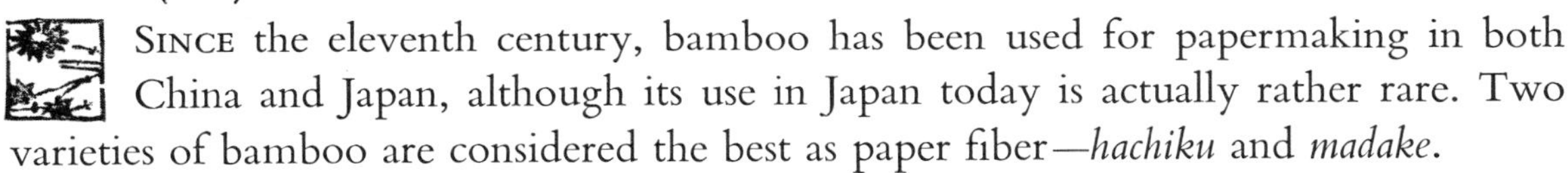

SINCE the eleventh century, bamboo has been used for papermaking in both China and Japan, although its use in Japan today is actually rather rare. Two varieties of bamboo are considered the best as paper fiber—*hachiku* and *madake*.

Bamboo is good in *gasenshi* because it absorbs the ink nicely, lends a pleasing beige tone to the paper, and makes the paper supple. The best Chinese paper of old contains bamboo.

The fibers of the bamboo are rather short, although longer than straw. The papermaker smashes the bamboo with a hammer to break up the fibers, then boils the material in a strong alkali solution for about three times the time required for straw. Bamboo, straw, and miscanthus fibers have the advantage of spoiling at a slow rate after processing, so they are particularly exploited in warmer areas of Japan.

*Miscanthus* (*kaya*)

THIS is a kind of reed used to thatch farmhouse roofs in Japan and sometimes, though rarely, is also used as a filler for paper, to stretch out the fiber content. Its characteristics and processing are very similar to that of straw. It also is sometimes an ingredient in *gasenshi*. It is more commonly used to form the papermaking screen and other tools than as a paper fiber.

Other fibers of less importance include cotton, *kuwa* (white mulberry), and *kaname-kōzo* (Chinese hawthorn).

*Mucilage*

ESSENTIAL to the Japanese method of making paper is the mucilaginous substance present in the root of *tororo-aoi* and in the inner bark of *nori-utsugi*. When this viscous substance similar in appearance to raw egg white is added to the paper solution, the paper fibers float upward and evenly throughout the vat. The *neri*—a common term for this mucilage among hundreds of vernacular terms—prevents the fibers from entangling and clumping together while the sheets are being formed, so that thin, even sheets are possible. The *neri* also decelerates the speed of drainage through the mold. Moreover, it acts in some mysterious way to allow the wet sheets to be separated after they are piled and pressed together without the benefit of interleaving felts. Most probably this is because the *neri* helps bind the fibers to each other, thereby strengthening the paper.

Although *neri*, or *nori*, translates as "starch" or "glue," these terms are misleading—"mucilage binder" is closer to the real function. It has also been called "sizing," but bears no similarity to the function of size in Western papermaking.

*Tororo-aoi* (*Abelmoschus manihot*, Medicus)

*Tororo-aoi* is an annual plant of the genus *Abelmoschus* of the mallow family (*Malvaceae*). Long ago it was brought from China, and now grows both wild and domesticated all over the Japanese countryside. There are two varieties of *tororo*, one with pink flowers and one with yellow, the roots of the yellow bearing a better mucilage.

*Tororo* is sown just before the rice planting—in Kurodani that is about the eighty-eighth day from the beginning of spring according to the lunar calendar. In July, farmers pick off its front and lateral buds, leaving only the best flowers to produce seeds. This is done so that the root will grow fat—hence it is short and gnarled, resembling, my teacher writes, "a deformed dwarf." The roots are harvested sometime between November and early December, depending on the climate, but always before the first frost. The roots vary from eight to twelve inches long. Formerly papermakers hung them from the eaves to dry and used them as needed—usually running out of supply before the papermaking season was over. Today the roots are cleaned and immersed in a preservative solution that keeps the mucilage usable for about a year.

Other varieties of *tororo-aoi* and related plants used for the same purpose are: *Althaea officinalis*, L. (marshmallow); *Althaea rosa* (hollyhock); *Abelmoschus esculentus* (okra); *Hibiscus manihot*, L. (rose mallow).

*Nori-utsugi* (*Hydrangea floribunda*, Regel; *Hydrangea paniculata*, Sieb.)

*Nori-utsugi* is a deciduous tree of the saxifrage family (*Saxifragaceae*), which grows on hillsides throughout Japan and especially in Hokkaido. Some soft areas of its inner bark function in the same manner as the root substance of *tororo-aoi*. It is an important source of mucilage, but because the tree takes several years to grow, it is second to *tororo* in use.

*Nori-utsugi* grows to a height of about six to ten feet. Large hydrangea-type blossoms appear in July and August. The method of cultivation is interesting—the head of one long shoot is merely bent over into the ground, and roots form and grow. In harvesting, parts of the tree are chopped off about a foot from the ground and stored by standing them in a stream until they are processed. After the inner bark is removed, it is processed in much the same way as *tororo-aoi* and sometimes used in combination with the latter.

*Nori-utsugi* is reputedly preferable in the making of *hōsho*, a quality paper used for woodblock prints, since the paper's surface has a character more like hemp paper.

Some other mucilage plants mostly not in use today are: *kazuro-zana* (colloquial term from the southern islands), a kind of vine; *urihada-kaede* (*Acer rufinerve*, Sieb. et Zucc.; related to the maple); violet roots; the root of *manjushiyage* (*Lycoris radiata*, Herb.); and the seaweed *funori* (*Gloiopeltis furcata*).

*Techniques*

All true paper, whether it is hand- or machine-manufactured, is made basically the same way: raw fibers are separated from the plant or raw material; the fiber is reduced to its cellulosic content by heat and chemical process; the reduced material is beaten to soften, separate, and hydrate the fibers; the pulp is formed into sheets by a sieving process; the sheets are dried and finished.

Briefly, in the case of *kōzo*: tree shoots are harvested, and the bark steamed and stripped from the shoots. The black skin is scraped from the inner white bark; the latter is washed in running water, boiled in an alkali, washed again, and bleached in some manner. Then it is picked of small impurities. The fiber is beaten by hand or machine, and a stock pulp solution prepared. To this a vegetable mucilage is added. A screen is dipped into the solution, and the wet, newly laid sheets are stacked. The stack is pressed gradually of excess water, then the sheets are brushed onto boards and dried in the sun. The finished sheets are inspected, sorted, cut, and packaged for distribution.

The following is a detailed description of the method of making paper by hand in Japan. For reasons of simplification, the method of making plain *kōzo* paper will be described in the main. It is the most common type made, and the methods used are similar enough to other Japanese papers to give a sound understanding of how all washi is made.

*Planting*

FOR many centuries *kōzo* has grown wild all over Japan. Now the wild varieties are diminishing rapidly. *Kōzo* is easily cultivated and yields more bark of better quality in its domesticated state than in the wild. The sowers and harvesters are not necessarily papermakers, but rather farmers who grow crops for the diminishing number of paper craftsmen.

*Kōzo* can be grown in either wet or dry fields. Wet fields are terraced much like rice paddies, and the workers prefer planting wet in summer because the water keeps them cool. But in general the seedlings are planted in spring on a southern mountain slope—about two feet apart to give good sun and air. By fall they have grown to a height of from five to six and one-half feet, although it takes two to three years for the bark to thicken and mature sufficiently to be used in papermaking. The plants grow seven or more shoots annually—so after the fourth or fifth year, one tree's bark yield is comparatively high. Between its tenth and twentieth year, the tree's bark is no longer usable. New trees can be grown by transplanting shoots, which soon take root.

*Harvesting*

*Kōzo* can be harvested any time the leaves begin falling, which is sometime between November and December. In certain regions it is traditional for the harvesters to go into the fields and pick the small mushrooms that grow on the *kōzo* stalks; when the mushrooms taste ripe and delicious, it is time to cut down the *kōzo*. Experts say the best *kōzo* is harvested around the time of the winter solstice, but before the big frost comes. Bark harvested then is called *akikawa*, "autumn bark." Usually a second crop is harvested in February or March, known as *harukawa*, "spring bark."

The shoots are cut close to the base and trimmed to a uniform length from between 40 to 120 centimeters (fifteen inches to four feet), all depending upon the size of the

papermaker's steamer. The whole tree is not cut down—only the shoots are cut off, and in another year the tree will have grown new shoots. Harvesters tie the shoots into bundles of manageable size, about twenty kilograms (about forty-five pounds) each, and, balancing two bundles on either shoulder, haul them off to the area set up for steaming. A good worker can cut twenty bundles a day. In areas where *kōzo* grows on mountainsides, this is hard work indeed.

*Harvesting Mitsumata and Gampi*

*Mitsumata* seeds are sown on hillsides as early as June. They are planted between young Japanese cedar and cypress trees, where the *mitsumata* helps shelter the slower growing cedars from the wind. *Mitsumata* does not yield good bark until it is three years old. The farmer is able to harvest matured shoots in late fall. *Mitsumata's* beautiful white flowers fall and fertilize the field around it. The buds burst in December, and flowers start blooming at the beginning of March, just before the leaves. This is the time of the second harvest. The shoots are cut several inches above ground and bound into faggots. The bundles are placed upright in a river to preserve their freshness until the papermaker is ready to steam them.

*Gampi*, as stated earlier, has always eluded large-scale cultivation. The wild plants are harvested in spring, from February to May, when the sap is rising and when the plant contains the most liquid. At this time only is it possible to peel the bark from the shoots, since *gampi* cannot be steamed like the other fibers. The peeled bark is dried in the sun and thereafter processed in a manner similar to *kōzo* and *mitsumata*.

*Steaming*

SINCE only the bark of the tree is used for making paper, it must first be separated from the shoot. In the case of *kōzo* and *mitsumata*, steaming loosens the fiber from the woody stalk. This process usually takes place the same time the shoots are harvested; the steamer may be located indoors or out. Since many hands are needed, it is usually a community project.

There is more than one method of steaming. In the more common one, the shoots, bound and cut to about three-foot lengths, are set upright and hammered tightly into a flat-bottomed iron cauldron, shoots extending above the cauldron. The size of the cauldron differs from one locality to another, but an average size is three feet in diameter

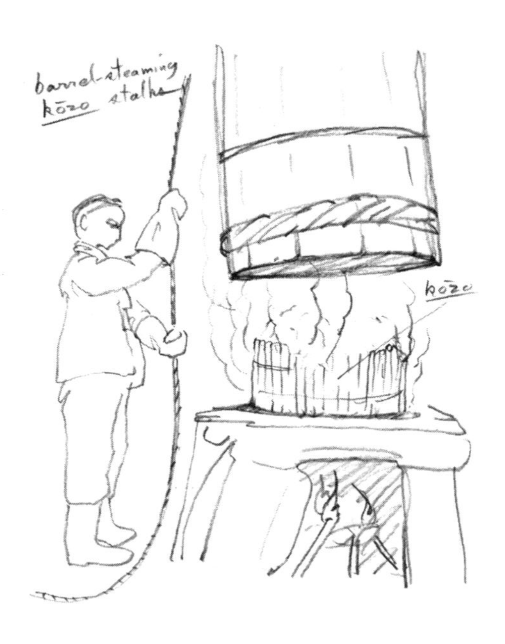

and about fifteen inches deep. A woven straw ring is usually placed over the top edge, acting as an insulator, and then a wooden tub or barrel of slightly greater diameter is placed over the shoots and pot, covering the exposed shoot tops. A fire is lighted under the cauldron and stoked continually for about two hours, or until the bark begins to separate from the stalk. Then the wooden top is carefully removed and cold water immediately poured over the shoots.

As the men remove the shoots from the pot, the women and children start right in, for the bark is easier to strip from the shoots while hot. Bark steamed the night before can wait until morning to be peeled, but it is usually done immediately after steaming, when the bark is moist and supple.

Usually all the local farmers or papermakers and their families gather for several days of concerted effort at harvest time to finish the steaming. This is considered the most lighthearted time for the papermakers, marking the beginning of a season of papermaking, and the atmosphere is as gay as a summer picnic.

A second method I observed employed a tall, narrow metal shed or closet constructed on the bank of a river. Close by were fields where *mitsumata* was being harvested. Cut shoots stood upright in the river. Water was brought from the river to feed the steamer. Under the shed, below ground level, lay a cauldron of water covered with a floor of wooden slats, allowing steam to rise into the closet.

Several men worked together packing the shoots into the steamer—the first several bundles lay horizontally across the bottom to utilize maximum space, and the rest stood upright. It took three men pushing and stuffing to crush the door closed and bolt it; rags were stuffed into the door cracks at the last minute.

Shoots were steamed this way for two and one-half hours, one person constantly stoking the fire. When the time was up, a steam whistle ingeniously rigged on top of the shed shrieked over the countryside, and the farmers came running to unload and strip the bark. When the door was opened, a shower of white-tufted *mitsumata* flowers tumbled out in a burst of steam.

*Peeling the Bark*

Most of the peeling is done by the women, often with the children's help. Workers kneel or sit in the winter sun on straw bags or mats laid over the ground. With her thumbnail, a woman makes an incision through the bark at the base of the shoot, twists the bark sharply, and the whole skin strips off in one piece about half-way up the shoot—or as far as she can reach, if the shoots are long. After stripping several shoots in this way, she ties the loose ends of bark together with one of the bark skins and places the bundles aside. When they have accumulated a sizable number of these bundles, the women and children group into teams of two—one grasps the tied bark ends, the other the bare end of the shoot and, uttering an encouraging cry, tugs strongly and evenly. The bark rips off the shoot.

The denuded shoots are often given in payment to the owner of the cauldron for winter fuel, but the women each keep aside a few of the cream-colored sticks for their flower arrangements.

The peeled bark is then dried in the winter sun, hanging from wood or bamboo poles. After the initial drying, the papermakers may suspend the bark from the eaves of the houses or even lay it across snow-free roofs. Drying takes several days. Then the stiff bark is tied into bundles of ten-*kan* (about forty kilograms) units, and may be trucked to papermakers in other areas or stored in storage houses indefinitely. At this stage the bark will not rot.

Such bark is called *kurokawa*, or "black bark," because of its thin, dark brown outer layer. After separating this from the inner layer, called *shirokawa*, or "white bark," the latter will be made into paper, and the black bark discarded or used in the manufacture of *chiri-gami* (see p. 173).

*Removing the Black Bark*

As winter advances, the papermaker takes his black bark from storage. This must be well softened with water before the dark outer layer can easily be removed. The bark is taken down to the river and, weighted down with a stone or floated in a dammed-up area, allowed to soak. In summer the bark absorbs sufficient water in about

half a day, but in winter it requires a full day or even overnight in the water. When the bark is adequately soaked, the thin, dark outer skin becomes flaky and loose.

In a very old and very effective process to begin removing this dark layer, the craftsmen tread on the bark in the river. It is placed on a flat stone in a shallow part of the water and trodden upon for a quarter of an hour, much of the dark skin flaking off and washing downstream. In the old days the papermakers trod on the bark in their bare feet, even in winter when the stream was half frozen. Today, if they use this treading process, they wear rubber boots.

The wet bark, trampled upon or not, is placed on a wooden board and scraped along the grain of the fiber with a wide-bladed knife. This removes the remaining black bark as well as some of the sticky green skin underneath it. The more black and green skin removed, the whiter the paper. Almost invariably the bark bears scars and rough areas made by insects or disease, so these, too, must carefully be cut away with the knife. Scrapers differ from one area to another, as do the boards on which the bark is scraped. In the far north and northeast, the flat top of a log is surrounded by a nest of braided straw—an extremely colorful tool.

This task is often done by women, children, or older members of the family because it is monotonous and not very strenuous work. There is always bark to be scraped, and this is one of the most time-consuming steps of all in the papermaking process. A conscientious scraper can clean about two *kan* (7.5 kilograms) of dry black bark in a day.

Tadao Endō, the fine papermaker of Shiroishi, Miyagi Prefecture, has invented a machine that takes much of the time and strain out of bark scraping. His machine does a very respectable job—but then his *kōzo* is of excellent quality to start with, being broad and largely free of knots. Endō's machine is, however, the exception, for this bark scraping is usually a slow and tedious hand process.

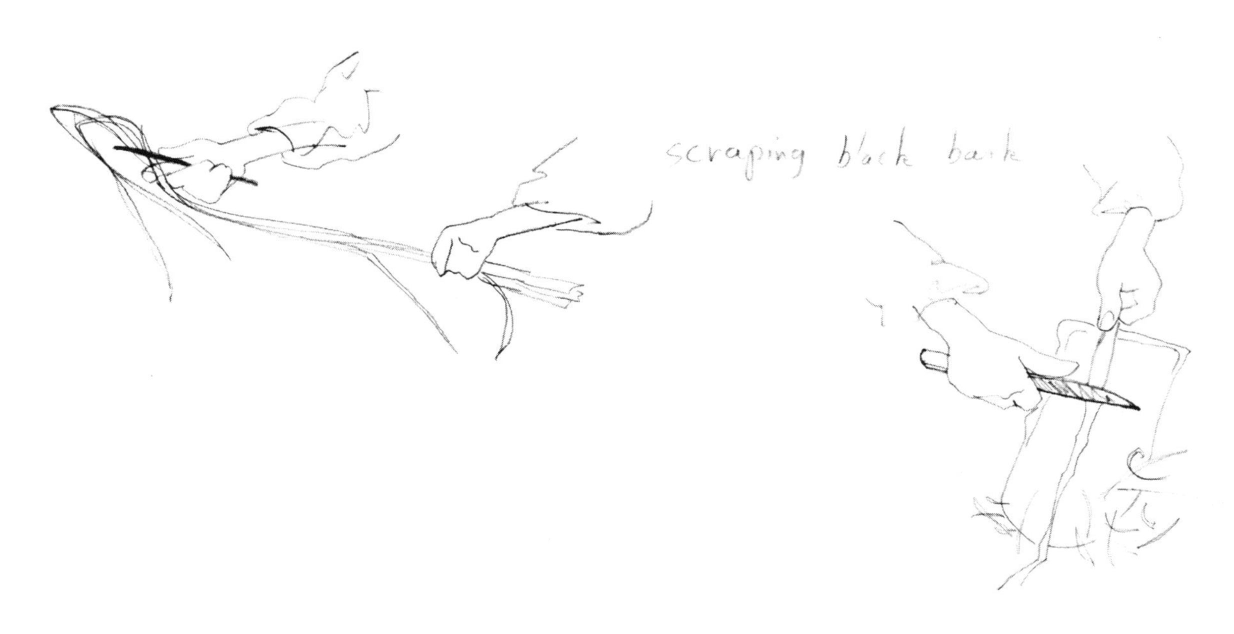


The flakes of black bark and especially the stringy fibers so carefully removed from the bark are often kept to use later in making *chiri-gami,* literally "refuse paper." This coarse, gray paper was the toilet paper of old Japan. Actually *chiri-gami* can be the most beautiful of papers. My teacher, Mr. Gotō, once said that a papermaker's integrity is reflected in whether or not he makes this paper from the dregs of black bark. After producing a fine paper from the inner white bark, a good papermaker takes these dregs, ferments the matter, beats it, and adds this *chiri* fiber to the last of his white paper solution. This was part of the natural Japanese parsimony—nothing should be wasted. Worst of all is the lazy papermaker who does not spend enough time and trouble to scrape off the black bark, but counts on harsh chemicals to do the job as the fiber is boiled. Modern variations of *chiri-gami* can be extraordinarily beautiful. Just for texture and interest, a papermaker may take some thin flakes of black bark and throw them into his vat of pure white paper, as does Eishirō Abe.

Not only the black bark, but also the green inner film, or cuticle, of the bark must be removed to obtain really white paper. Mr. Gotō's paper differs in tone from most every other paper I have seen because he leaves on some of this green skin. The residual film imparts a distinctly natural color to his paper—a hint of green as well as a sweet vegetable fragrance. Most of the cuticle comes out in the boiling and scraping, and more when it is well washed. All the wounds, scars, and undesirable parts of the bark should be removed at this stage, because after the fiber is boiled and beaten it is extremely tough to pick out.

The white bark bereft of its most conspicuous impurities is rinsed thoroughly in a stream to remove the last of the glutinous matter. Then it is dried in the sun over bamboo racks for two to four days. Drying the cleaned bark in the sun and on top of the snow in winter helps whiten it. At this stage it is called *shirokawa,* "white bark." Like *kurokawa,* it can be stored indefinitely until it is ready to be processed further. This,

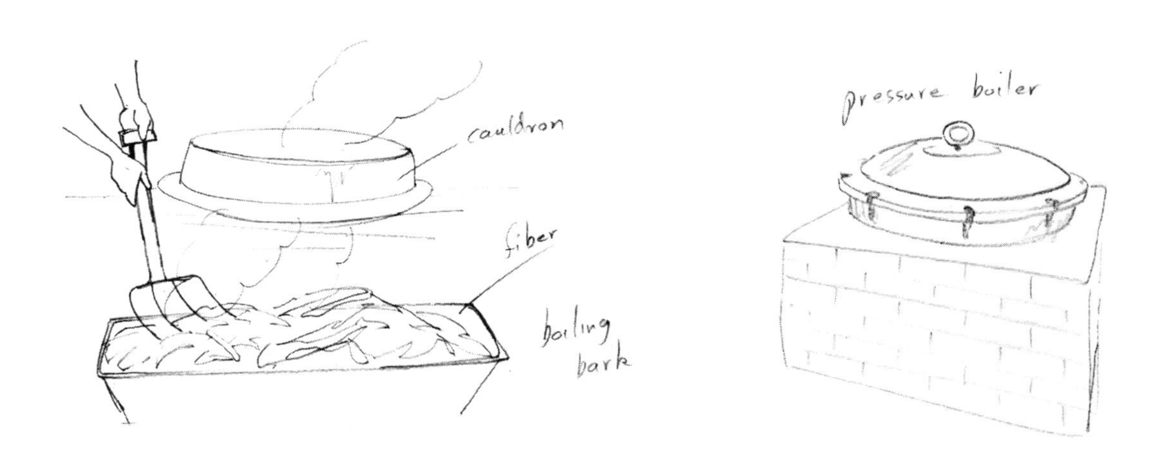


however, is the last stage at which the fiber can be stored for long periods; once it has been boiled, the processing must continue uninterrupted until the paper is finished. Boiled fiber is easily susceptible to mildew, especially during the humid Japanese summer months.

*Boiling the Fiber*

All steps in the manufacture of Japanese handmade paper depend much more on experience and intuition than set methodology, but especially so in the process of boiling the fiber with an alkali. It is much like country cooking but on a larger scale—a tubful of this, a cauldron of that, and a lot left up to nature.

Basically, a fire is started beneath a cauldron of water, the white bark is added, and, after a period of boiling, an alkali is added. This boiling process softens the fiber considerably, and the noncellulosic material in the fiber is digested—starches, fats, and tannins in the fiber are released. Throughout Japan, temperature, pressure, and cooking time vary. All are more or less effective. Traditionally lye or buckwheat ash or other available, mild alkalis were used. But in the past sixty years or so much harsher chemicals such as caustic soda have been employed because they are cheaper, easier to rinse away, and because they digest impurities well. However, because such chemicals also weaken the fiber considerably, many purists will have nothing to do with them.

Before the boiling begins, the *shirokawa* is presoaked in water for about twelve hours. This whitens the fiber further and makes it easier to boil. Papermakers without a river nearby place the fiber in a tub, but clear, flowing water is best. Meanwhile a large cauldron of hot water is prepared. The soaked bark is tossed into the pot just before it begins boiling, and an alkali is added. The type of alkali and how much of it is used influences greatly the quality of the paper. Most papermakers use soda ash, slaked lime, or the harsher caustic soda. Before the days of bottled gas, kerosene, and electric stoves, papermakers had their own hearths as a good source of wood ash, which, all agree, makes the best lye. Living National Treasure Ichibei Iwano used buckwheat ash.

Alkali is added in the amount of about one-tenth the weight of dry white fiber. More is used for black bark fiber. The solution is boiled for about one hour, sometimes two, or until the papermaker can crush the fiber between his fingers. Then the fire is extinguished and the covered pot allowed to sit anywhere between one and eight hours.

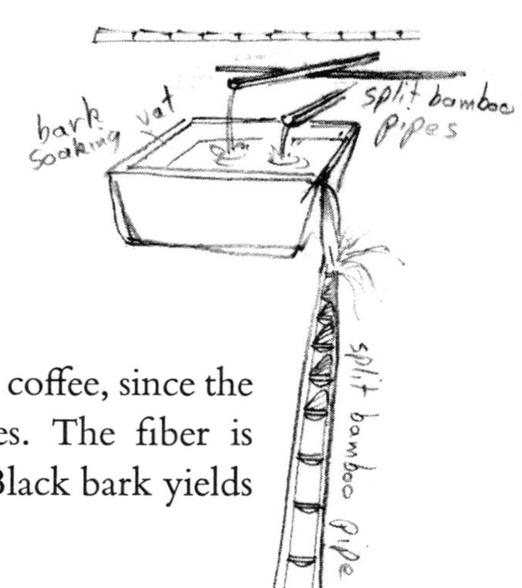


The solution in the pot will have turned into something resembling coffee, since the chemicals break down the fiber's fats, starches, and astringent juices. The fiber is removed from the pot, put into a large basket, and rinsed thoroughly. Black bark yields about 50 percent usable fiber; white bark, 80 percent.

*Bleaching the Fiber*

The drained fiber is carried down to the river. Small bundles of the *kōzo* are bound together with one end of the fiber, or put into baskets, and immersed in a dammed-up area of the stream. Here the fiber lies for half a day in summer or at least two days in winter. This is called *kawa-zarashi*, "stream bleaching," since the fibers begin to turn very white as the last of the alkali washes away. In some villages in Shikoku, rivers rich in limestone minerals whiten the bark further. In Kuzu, Nara Prefecture, the bunches of floating bark are a beautiful sight—white stars suspended in deep valley rivers.

Stream bleaching is a very important and effective step in the making of fine washi; the fiber is not weakened—some say its strength is even enhanced—and the brightness of the bark is soft and natural in tone. But like all the finest traditional papermaking methods, *kawa-zarashi* is disappearing. One reason is the serious pollution ruining Japan's beautiful rivers. But a larger portion of the blame must be put on the paper-makers themselves, who use chemical bleaches because they are cheap, quick, and readily available even though they weaken paper fibers. Of these bleaching agents, bleaching powder is the most common. Unfortunately it is sometimes used in villages where even the best washi is produced—although only for papers that must be very white and not overly durable. It seems a shame and a waste that hand paper craftsmen feel they must resort to the use of such chemicals.

Two other methods of bleaching that take advantage of natural conditions are *yuki-zarashi*, "snow bleaching" and *tenpi-zarashi*, "sun bleaching." These methods seem to be used rarely today. Snow bleaching is sometimes still seen in places such as Toyama and Niigata prefectures, where snow falls heavily but where there is little running water. The fiber is laid upon the snow and partially buried in it. In sun bleaching, it is simply laid upon the grass and allowed to soak up sun.

*Picking Out the Dross*

Considered the most tedious process of all in the making of washi is picking out the scars and the residual matter that has still stubbornly clung to the fiber. This is *chiri-tori*. Often a small sluiceway is built from cement in a semi-enclosed area, and

water is channeled from a nearby river or tap. The workers kneel or sit by the sluice or dammed-up area of the river and immerse their big baskets of fiber in the water. As water flows through the baskets, fibers float to the top, where they are picked clean with fingernails or a needle. Clean, cold water running through the bark helps further whiten it. Scarred, discolored, and decayed bark is thrown aside. When the bark is sparkling clean and snow white, it is rolled into melon-sized balls and wrung of excess water.

For the best-quality paper, the fiber will be washed over and over again after this. In some localities small, portable thatched huts were brought down to the river beside a dammed-up area. There in sun, wind, rain, and snow the men and women sat washing the white bark over and over again in the clear water. Such washing whitens the paper naturally without the use of chemicals.

*Beating the Fiber*

THE purpose of beating the fiber is to break apart the individual fibers, shorten, soften, and hydrate them so that a close, even sheet of paper is possible. Before the introduction of beating machinery about one hundred years ago, all beating was done by hand. Even while great mechanical beaters run with water power beat the rags used in the paper of medieval Europe, such machinery was unknown to isolated Japan. The very first Chinese papers were beaten in a large, primitive mortar and pounded with a wooden pestle; for a short while in her early days of papermaking this method was used in Japan as well. From then until very recent times entire papermaking families were involved in beating the bast with thick rods or bats or mallets on a hard surface, and in some isolated areas and in a very few quality mills today this old method is still practiced. Because it is traditional and still in use, and because it greatly improves the strength and durability of paper, the hand method of beating deserves description.

A flat, smooth stone, such as granite, or a thick hardwood slab serves as the beating platform. A melon-sized ball of fiber is placed upon it. The papermaker stands or kneels before the platform and beats the fiber with a hardwood rod. The shape, size, and length of these rods differ from one region to another, but most are about three feet long with squared-off edges. In some places such as Mino, large wooden mallets are used instead; the heads of the mallets are grooved with sunburst patterns. Every now and then a little water is thrown on the bast, and the fiber is hydrated as it is beaten.

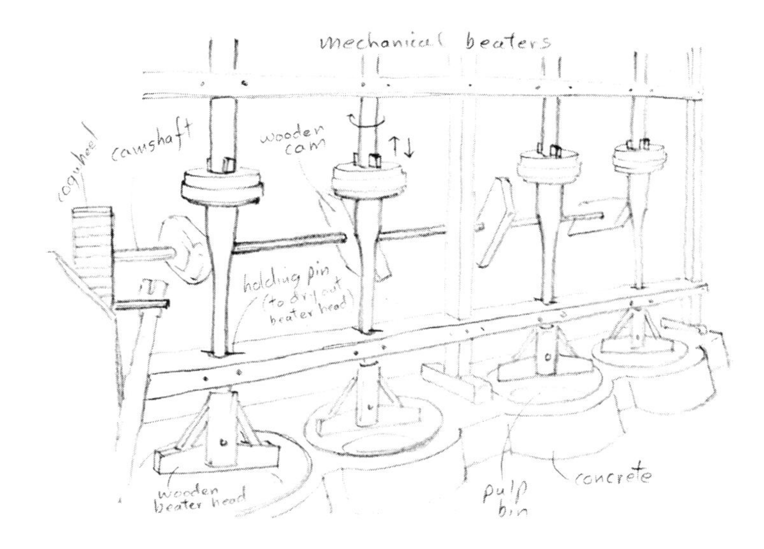


In days long gone, the family spent the day processing the fiber and stayed up all night beating it—quite laborious work. Often the members of the family knelt facing each other across the platform, working together at their beating. One old woman related that as a girl hand-beating the pulp, the early spring wind tore across her hands so hard the skin broke and bled. The steady, rhythmic, percussive sounds created by this monotonous pounding gave rise to many papermaking songs, some of which lament the fate of girls born into papermaking families and destined to work a full day followed by a full night's beating:

> The papermaking girls of Kuzu are worse off
> than beggars;
> Don't beggars sleep at night and in the
> daytime too?[2]

Today paper is rarely beaten by hand. Ichibei Iwano is the only papermaker I know of who hand-beats all of his fiber. Most papermakers employ a small Hollander-type beater, electrically powered, which is a miniature version of that used in machine paper manufacture. The Hollander is an oval tub holding water mixed with the processed pulp stock. The bast rotates around the tub and through an iron roller fixed with dull steel blades. The idea behind the Hollander seems to be to cut the fibers short and break them apart so a more even sheet is easily obtained. Machine beaters cut the fiber considerably both lengthwise and crosswise. Traditional hand-beating crushes the fibers to draw out the fiber ends, separates the single fibers, and preserves the fiber strength. This is most evident when observing both kinds of beaten fiber under a microscope—the machine-beaten fibers appear short with clean-cut ends; the hand-beaten ones are long with splayed ends.

But the machine has won. Most papermakers feel they cannot afford to spend the time it takes to beat all their fiber by hand. In Kurodani, electrically driven machine-beating is preceded by beating the fiber in a system of huge wooden hammers run by electric power, which resemble early European and American beaters. Sometimes a combination of beating techniques is used. For example, Seikichirō Gotō will often

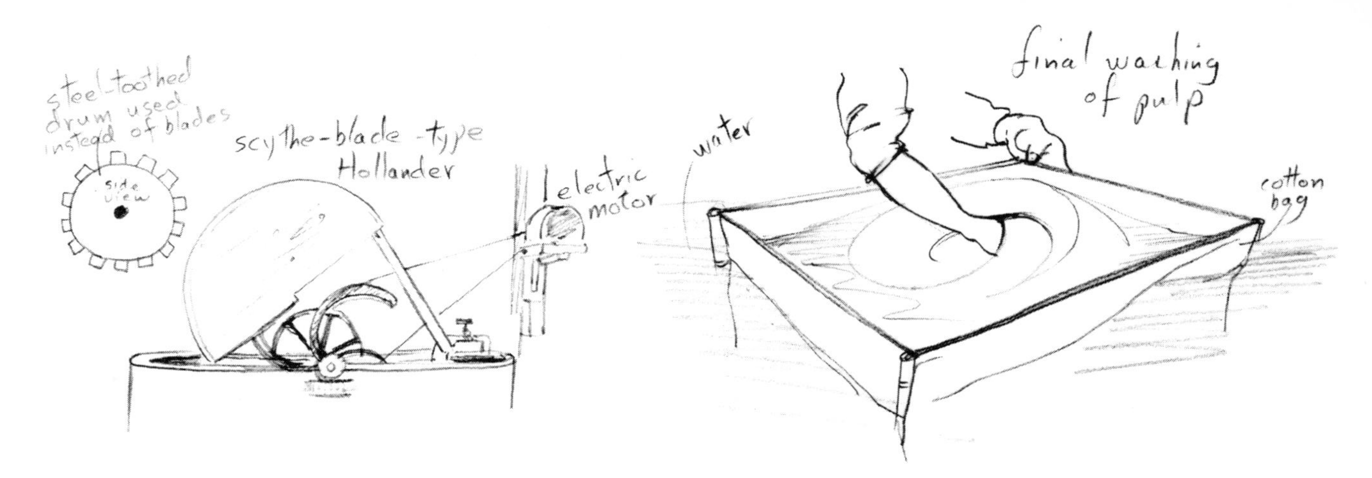


hand-beat a manageable quantity of fiber for about ten minutes, put the fiber through a mechanical stamper beater for another fifteen or twenty minutes, then complete the maceration in his Hollander-type machine.

The beating is completed when the fibers resemble wet cotton, breaking into single strands rather than clumps in water. It is possible to overbeat the fiber, which is not good, upsetting the vital water-cellulose relationship in the finished sheet. Additionally, the strands will tend to float too much on the surface of the solution while the paper is being made, making the vatman's task more difficult. For very high-quality papers, the fiber may again be washed, although in a fine cotton bag so that the strands do not escape. This is the final whitening and strengthening of the fiber, since fiber contracts in cold water.

### *Making the Mucilage*

THE mucilage, most often referred to as *neri* ("starch"), is the magic ingredient in the Japanese papermaking process—without it washi could not be made. This vegetable mucilage, besides allowing the fibers to float evenly in the vat, controls the rate at which they filter through the screen and makes it possible to separate the wet, stacked-up sheets without the use of felts. Surprisingly, the *neri* leaves no trace of itself in the finished papers.

Two parts of two plants described earlier are used in manufacturing this vegetable mucilage; the inner bark of the *nori-utsugi* tree and, much more commonly, the root of the *tororo-aoi* plant. Their processing is similar, but because the *tororo* is used most frequently, its processing will be described.

After being harvested in fall, the *tororo* roots are kept in a huge ceramic pot containing a phenolic preservative solution. This solution preserves the freshness of the roots for up to a year, when more roots can be harvested. At all times the undried roots are kept as cold as possible, just short of freezing. When the papermaker is ready to prepare the stock, he removes a root or two from the preservative, washes it well in water, then crushes it with a wooden mallet on a flat stone. In order for its mucilage to be released slowly, the smashed root is placed in a barrel with water and allowed to sit overnight or for a period of hours. Then the papermaker stirs the solution with a bamboo pole, mixing with water the mucilage that has oozed from the root. The *neri* solution resem-

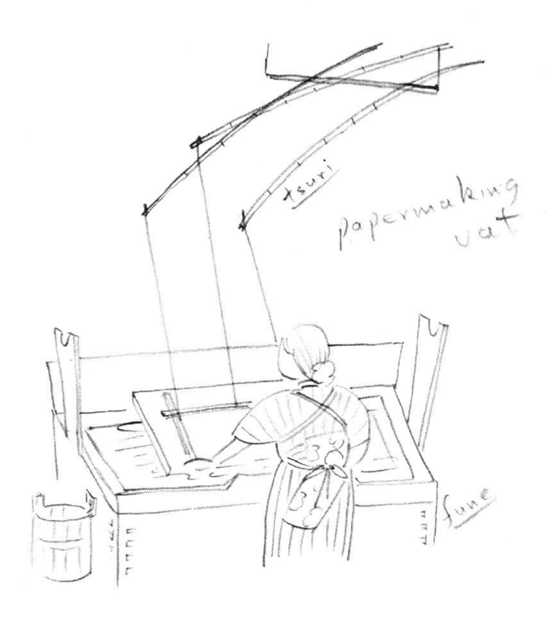


bles diluted egg white, clear and somewhat viscous. If more mucilage is needed, the root must be mashed some with the pole in the barrel or bucket. One root is exhausted of mucilage in about one to two days of papermaking.

The papermaker knows by a particular blurping sound created as he stirs the *neri* solution just when it is ready. Then the mucilage is poured into a cotton cloth bag held over the vat. The thick *neri* solution is squeezed through the bag into the paper pulp solution, the bag filtering out unwanted root particles. Some craftsmen employ a device similar to a small wine press to squeeze the *tororo* root of mucilage, with a cloth bag at the spout for a filter.

*The Workshop*

THE washi is scooped in a wooden vat called the *kamisukibune*, literally "papermaking boat"; or it is simply called the *fune*, "boat." Usually the vat is of maple and remains watertight with constant use. Sizes differ greatly, depending, of course, on the size of the largest mold used in the vat, but an average vat measures about 60 centimeters (two feet) or more across, 120 centimeters (four feet) long and 45 centimeters (one and one-half feet) deep.

The vat may or may not contain a mechanical agitator—"hog" in paper terminology—called a *sukuryūshiki*. Electrically powered, this harrowlike device mixes the fiber evenly in the solution. The papermaker stops from time to time and reagitates the solution with this machine. Instead of an electric agitator, many papermakers prefer to use a large wooden comb called *mase* or *zaburi*, which they operate manually.

Sticks lie across either side end of the *fune*. When the paper is made, the sticks are slid to the middle of the vat and used to support the resting mold. Commonly cords hang from the end of supple bamboo poles fixed in the ceiling rafters—the *tsuri*. When the cords are hooked onto the mold, the bamboos act as springs, easing the great weight of the paper solution as it is tossed on the screen. In front of the *fune* is a platform of wooden slats on which the papermaker stands, protecting his feet from pools of dripped water. Behind the papermaker, or at his right side, stands a bench topped with a board on which the finished sheets are stacked. Sticks or dowels hammered into the front corners of the board act as guides so that the freshly laid sheets will be placed squarely one upon the other. Close by are barrels of prepared mucilage and processed fiber. And behind the worker or to his side is a coal-heated hand warmer—a small wooden container filled with hot water into which he occasionally plunges his red, numb hands.

### *Making the Paper Stock*

To MAKE paper, the worker fills the *fune* about three-fifths full of very cold water and throws in wet balls of fiber at an approximate ratio of one kilogram of fiber to one hundred liters of water; this varies according to the type of paper being made. After a preliminary stirring with a pole, the papermaker briskly agitates the mixture with the *mase* or with the electric agitator until the fibers are floating evenly in the vat. Then the *neri* is filtered into the solution, and the solution stirred well with a bamboo pole. The veteran papermaker knows by experience and feel just when enough *neri* has been added. The most important tip-off is the sound—when he hears a deep, churning hollow noise as he stirs the *neri* into the solution, he knows the amount of *neri* is correct. My teacher described this sound as *chon-chón-chon-chón*. If you take a corner of a piece of good-quality, thick, crisp *kōzo* paper between your fingers and snap the paper through the center, you will hear just this *chon-chón*.

The more *neri* added, the more slowly the paper solution flows through the papermaking screen. Without *neri* or with too little, the liquid runs through the screen too quickly, allowing no time for the papermaker to form an even sheet and leaving clumps of fiber on top of the screen. If too much *neri* is added, the papermaker wastes time waiting for water to sieve through the screen. Also, I have heard, too much *neri* makes the nap of the paper stand up. It is the use of this vegetable mucilage that makes it possible to produce by hand extremely thin, fine sheets of great strength.

### *Making the Paper*

THE stock solution prepared, the papermaker places the screen into the frame. Together these comprise the mold. The sizes of molds differ considerably, depending on the type of paper being made and local custom and preference. An average mold used to make common *kōzo* paper is nearly one-half meter (eighteen inches) across and just over one meter (three and one-half feet) wide. One used in Mt. Kōya is only twenty-five centimeters (ten inches) across and thirty centimeters (one foot) wide. Some are so large that two or even four papermakers are needed to work with

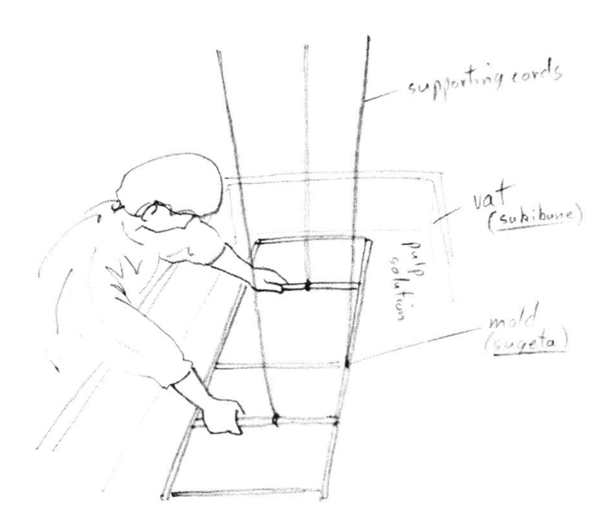


them, although such molds are highly unusual. Some frames are compartmentalized so that two or more smaller sheets can be made at the same time in one mold.

The screen, called *su*, is made of finely split and beveled bamboo splints woven together with silk threads—a remarkable piece of craftsmanship. In some areas, miscanthus and other reeds woven with horsehair are used, leaving faint characteristic "chain-lines" in the finished sheets. The screen is highly flexible, rigid on the weft and pliant on the warp, rather like a huge, finely crafted bamboo place mat. The worker clamps the *su* into the lightweight wooden frame, which has some degree of natural water resistance and is equipped with copper hinges and catches. This frame, or *keta*, has two handles on top spaced for the papermaker to grasp comfortably. Once the screen is secured into the frame, the mold (*sugeta*) is rigid.

The papermaker grasps the handles of the mold and plunges it into the vat, the side of the frame closest to him touching the solution first. He brings the mold down through the liquid and, mold horizontal again, scoops solution up out of the vat. The shallow edges of the frame serve to hold in sufficient solution. As the liquid slowly filters down through the screen, the papermaker tosses the solution repeatedly back and forth and from side to side, evenly and rhythmically, forming an even film of pulpy fiber on the screen. This tossing is called the "vatman's stroke." The solution is heavy—it takes not a small amount of expertise to hold, balance, and dexterously "stroke" the solution over the full mold with a slight shaking motion. By this shaking and stroking, the long fibers cross and intertwine, strengthening the sheet in both directions, and more water is forced through the screen. *Neri* makes this prolonged shaking possible—one secret to the great strength of washi.

Just before all the liquids have drained through, the papermaker tips the mold back so that the liquid falls to the rear of the screen, gives a deft flick of the wrists, and tosses the remaining solution out over the back edge of the mold into the vat. This casting off of solution—made possible only by the use of the mucilage—is one main distinguishing characteristic of Japanese hand papermaking. If wrongly executed, the sheet collapses. The discharge of the last solution allows the papermaker to get rid of accumulated fiber clusters and other matter; it also speeds up the process, since, as the sheet thickens, there is a slowdown in the speed at which water filters out. Some papermakers say that an

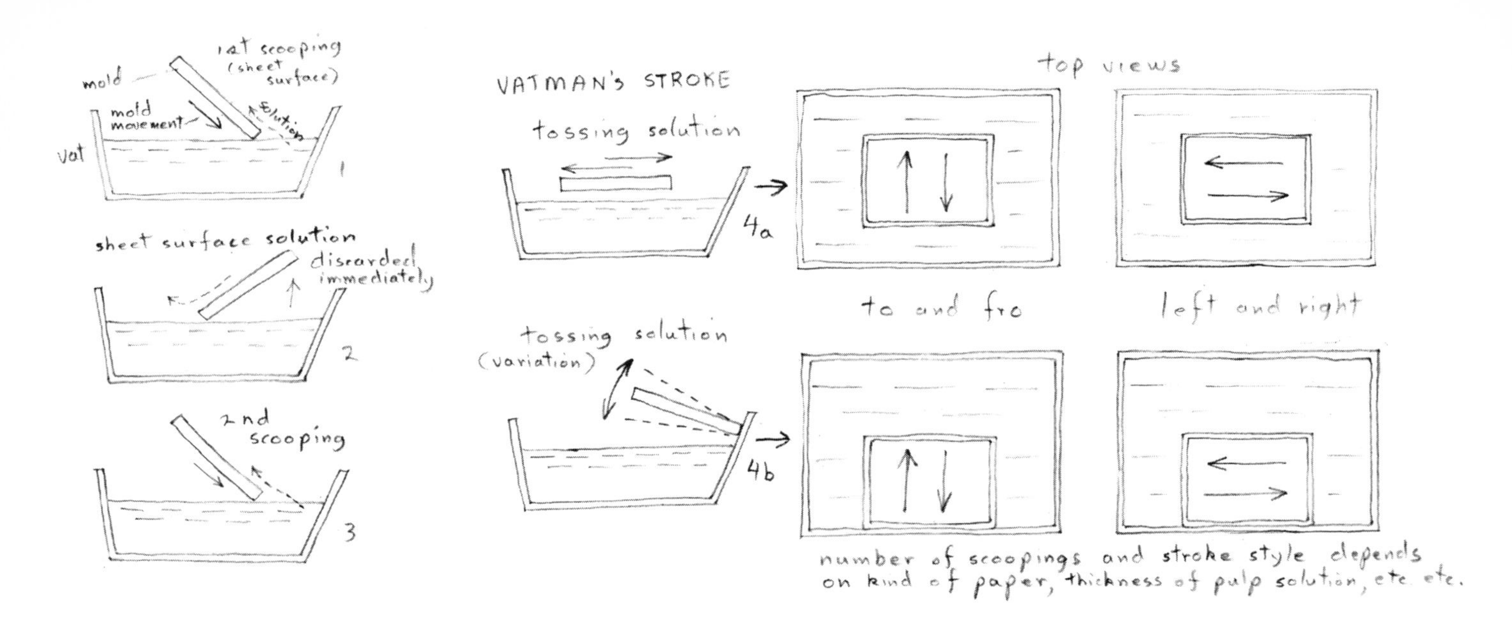


important feature of the stroke is the manner in which the thumbs are used. In the *nagashi-zuki* method, when the mold is dipped, the thumbs do not touch it; but once the shaking begins, the thumbs are pressed upon the mold for strength and stability. Such techniques as this may differ slightly from one area in Japan to another. The worker repeats the scooping, shaking, and discharging of remaining solution at least one more time, generally several times, until he attains the desired sheet thickness. As the stock gets thinner, more fiber, mucilage, and sometimes water are added. In this manner the sheets are formed.

It takes an extremely experienced eye to judge the correct thickness of a sheet each time as it is being made. Thin sheets of consistent quality are even more difficult to produce than thicker ones. Actually it is impossible to achieve precise thickness each time; but as in anything made by hand, precision is not the point. If it were, papermakers long ago would have devised some mechanical way in which to scoop the sheets.

To get the rhythm of the scooping process takes from two to four years in the making of large sheets, or about ten thousand sheets. Some say it takes ten years. Bunshō Jugaku has said that it takes three years to learn the correct hip rhythm and two to get down the neck rhythm—it is that difficult. Intuition and "feel" have much to do with the learning of this very important process, and some seem to have more talent for it than others. In Japanese this is *kyō gai betsu den*—"the most important thing cannot be taught"; it has meaning on many levels, even within this one craft. Sometimes, in the studio of a very fine papermaker, one can hear an almost musical sound as the worker sloshes the solution over the mold—"like the waves of the sea" someone has said. That particular sound represents solution at the right temperature and with the correct amount of *neri* moved across the screen by an experienced papermaker in tune with the rhythm of his craft.

Paper experts have drawn a distinction between two methods of hand paper manufacture: the method probably developed in, and for a long time peculiar to, Japan described above, called *nagashi-zuki*—"discharge papermaking"; and the method

originating in China that spread to Korea, Japan, and the West, called *tame-zuki*—"accumulation papermaking"—generally better for thick papers.

Western handcrafted papers are made the *tame-zuki* way. No vegetable mucilage is used; therefore the sheets are made quickly and with less shaking to and fro, since the liquids flow rapidly through the screen. The tossing is also less vigorous than in *nagashi-zuki*. Usually in Western *tame-zuki* the vatman will not discharge any excess solution at the end of his stroke, but according to Dard Hunter he sometimes will discharge a small amount, depending entirely on a papermaker's personal style. The sheets are put aside to drain for some time while still on the mold. After that they are cast from the screen onto a piece of felting and so stacked up, each sheet divided by felts. *Tame-zuki* is the method thought to have been brought to Japan in the seventh century, when papermaking was first introduced there. It is a technique still in use, though rarely, for the manufacture of certain papers, notably *maniai-shi* (see p. 182) and *Kōya-shi*. Moreover, a combination of the two techniques is also practiced, as in the making of *Nishinouchi-shi* (see p. 185).

The following is Sōetsu Yanagi's description of the *tame-zuki* and *nagashi-zuki* methods from his essay "The Beauty of Washi":

> *Tame-zuki* and *nagashi-zuki*—these are the two ways of making paper today. Originally they were one, but as time passed they became divided. The idea in *tame-zuki* is static; in *nagashi-zuki* it is dynamic. Both contribute to the formation of the world of washi. The former gently amasses the fibers and aims at thickness. The water quickly drips through the bottom, and a layer of paper is left. The famous *torinoko* of Echizen is made by this process.
>
> However, *tame-zuki* is not a method limited to Japan. The method of making washi that surprises people is the process of *nagashi*. As the liquid mixture flows over the papermaking screen, the fibers are lined up, matted, and piled up in whatever direction the hands that shake the screen move. At the moment of appropriate thickness, the process comes to an end with the consummate act of the tossing out of the liquid. Everything is a miracle of hands. Without this skillful hand technique, *nagashi-zuki* could not exist. That is why the word *tesuki*, "hand-laid," is appropriate. Most of the famous washi types are made by this process.[3]

When a sheet of proper evenness and thickness has been formed, the worker draws the two slats over from the sides of the vat and rests the dripping mold upon them. He

unlatches the frame, removes the *su* with a thin, wet sheet upon it and, inverting it, places the screen, sheet side down, upon the stack of fresh, wet sheets. With slight pressure he jerks the flexible screen gently, pulling it from the paper. The sheet sticks to the pile of wet sheets. Then, placing the screen back into the frame, he scoops again. Each sheet is lowered neatly over the last. An experienced papermaker can make two hundred or more large sheets in a day.

Other techniques are also often used. Some workers place a string or thin reed of straw or miscanthus between each sheet, primarily as a counter but also to help separate them. For special papers where no "chain-lines" (left by the reed in the screen) are desired, a cloth of fine silk gauze called *sha* or *suki-sha* is placed over the screen before it is locked into the frame.

*Pressing*

THE stack of dripping sheets is covered with a cloth or with the papermaking screen and left overnight. The following morning the sheets are pressed of their excess water by one of two methods, traditional or mechanical. Either way, the sheets must be pressed slowly and carefully so as not to damage them. Pressing is as much an art as any other step in paper manufacture and must be executed with just as much patience and attention.

The traditional method makes use of a simple pile of stones or of a log acting as a lever. In the latter, a large log is anchored against a wall or wooden post, usually with stone weights. The board of stacked wet papers, or "post," is covered with a cloth and another board and placed under the fulcrum point of the log not far from the wall. Stones with wire loops are strung over the projecting end. Gradually more stones are added as the water slowly seeps from the stack of pressed sheets. Pressure is gauged purely according to experience. This method is considered by many the best. The other method employs screw or hydraulic jack presses, the more modern ones equipped with pressure gauges.

The wet paper requires about ten hours or more of gradual pressing. The *neri* forms an invisible coating, preventing the sheets from adhering even after pressure has

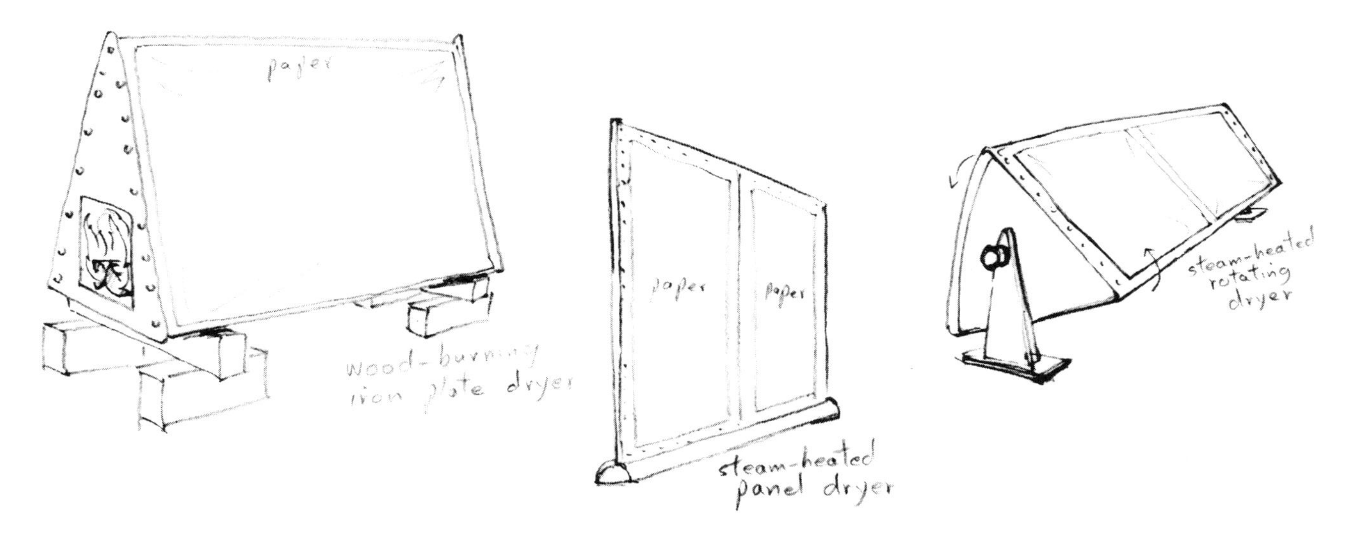


been applied. If the sheets are pressed too quickly, they will adhere beyond separation. If not pressed enough, they will retain so much water that their separation damages them.

*Drying the Sheets*

Even after being pressed, the sheets contain a high percentage of water—about 80 percent—just enough to allow them to be separated by hand. Usually the drying must commence twenty-four hours or so after pressing, depending on the moisture in the air. Drying paper on boards out-of-doors is the traditional method; but because weather is often fickle, paper is increasingly coming to be dried on iron heaters or in a drying room. Drying paper by sun and wind produces strong, bright, crisp sheets.

Sheet by sheet, the worker peels paper from the damp stack. Catching a needle or his fingernail in a corner, he pulls in line with the fibers or diagonally on the bias, being careful not to rip the sheet. Sometimes he uses a round bamboo dowel or tweezers to aid him. Finding the bottom of a sheet is not always easy, since a sheet can actually be divided into as many layers as the times the screen was lowered into the vat. Although the worker must take care, the sheets are amazingly strong even when wet.

Each sheet is laid upon a drying board, generally about two meters (roughly six feet) long by sixty centimeters (two feet) wide. Usually they are of pine, gingko, or horse chestnut, more rarely cherry, cryptomeria, or laurel. The surface of the board has been cleaned and planed smooth; but some old boards are no longer planed, because the relief of wood grain makes a lovely impression in the paper. Some boards still in use are hundreds of years old.

The papermaker presses the sheets onto the board with a wide, soft brush made of various materials. Horsehair and deerhair are the best, but palm fiber is also used, particularly in the southern islands. The worker brushes the damp sheet swiftly, with light, even strokes along the fiber grain and with infinite care—in a second he could destroy that into which he has put so much time and labor. He places the top surface of the sheet—called the *omote*—against the board. When the sheet is dry this surface will be smooth and the exposed, rough side, the *ura*, may bear faint brush marks.

Covering both sides of the board with paper, the worker carries the board into the sun, resting it against a support. The wood of the board helps absorb some of the paper's moisture. Eventually the yard or hillside is white with rows of drying sheets. In strong sun and wind the sheets dry quickly, in a matter of minutes or hours. A thousand or more sheets can thus be dried by one worker in one sunny day. The dried paper is removed with a bamboo spatula; then the boards are turned to expose the opposite side to sunlight.

On rainy or snowy days, papermakers often make use of indoor drying methods; otherwise, if it is not dried in two days, the paper would take on a dull, lusterless finish. In olden times, papermakers brought the boards indoors by a fire. A few such as Ichibei Iwano dry the paper on boards in a dry, heated room. More frequently today papermakers make use of sheet iron dryers heated from within by gas, oil, or wood.

Every papermaker admits that drying with iron dryers is the least desirable of all methods, though eminently practical. High temperature drying frustrates the natural shrinkage and weakens the fibers. Artificial drying makes the sheets all the same—"half-living, half-dead" as one man described it. The sheets do not dry from the core; the paper is flaccid.

Not frequently, but sometimes, paper is dried with two or more sheets unseparated on the board to save space. Such composite sheets may be used thick as they are or the components can be separated by peeling them apart from the corners.

### "*Finishing*"

By and large the Japanese do not add a "finish" to their paper. Washi was traditionally meant for use with brush and *sumi* ink, so sizing was not particularly desirable; however, artists and printers may size their paper with a mixture of bone glue and alum before using. Occasionally the papermaker may treat some sheets in this fashion.

It was once the custom to add some rice powder or ground seashell to the vat to impart a luster and whiteness to the paper. These as well as clays increased the paper's weight and thickness considerably—not a few papermakers added them to cheat the bulk buyer. Powdered minerals such as kaolin are still added to the vats of some special papers, such as *maniai-shi* (see p. 182) in Najio, near Kobe. They make the paper's surface smooth, almost greasy to the touch, and close the pores. These, however, are for papers with special functions; most papermakers neither add sizing (other than *neri*) nor calender the sheets, which they consider adulterating or extraneous.

*Inspecting, Cutting, and Packing*

THE papermaker, as he peels the sheets from the drying board, makes a quick preliminary judgment as to the quality of the sheets, dividing them accordingly into two or three piles. Then the sheets are brought into a clean room with good natural lighting and individually inspected. The inspector looks for holes or injured areas, uniform thickness and weight, and finally good color and complexion. A skilled person can inspect sheets with his hands and eyes alone. The poorest sheets, such as those made by a beginner, are recycled into new sheets—perhaps about ten percent of the total. Second- and third-quality sheets are sold at discount as "seconds."

Washi has the extraordinary attribute of increasing its strength and weight with time. In ancient days, paper was sold by weight, but since weight is so irregular in handmade paper—and since papermakers began to "load" their paper with heavy clays—now washi is sold by sheet units. Many sheets are neither cut nor trimmed of their natural deckle edge, but packed and distributed as they are. The cutting of paper is a distinct art. Expert cutters who make the rounds of the papermaking houses are sometimes employed for this task, but just as often the papermakers themselves learn how to master this skill. The manner of cutting and shape of the knife differ from one region to another; but generally the knife has a blade that curves inward. Some resemble sickles; others, cleavers. The knife blade must be extremely sharp.

In order to cut the paper according to its standard size, a Japanese carpenter's rule (*jōgi*) is used. The rule is usually no more than a board of cherry or *keyaki* wood. About two hundred sheets are placed on a bottom board, usually of willow, and the rule is placed on top of the paper. The cutter steadies the rule with his foot or knee. First the deckle edges are trimmed away and then the stack may be cut down the middle.

Washi is traditionally counted in units of *jō* and *soku*. One *soku* usually consists of ten *jō*, but the *jō* unit depends on the kind of paper and locality, since even the same kind of paper varies from one place of production to another. These units are highly confusing and inconsistent, differing for each paper. In the case of *hanshi*, generally one large sheet is cut into four pieces. Twenty large sheets (or eighty cut sheets) comprise one *jō*; ten *jō* one *soku*; and ten *soku* one *maru* or *shime*, depending on old or new measuring terminology. At any rate, stacks of ten *soku* are placed on top of each other to form a *maru* or *shime* and wrapped with three to five sheets of the same kind of paper, overlapping, to cover. The wrapped bundle is bound with string made of white bark *kōzo*, inspected, and sent to market. Papermakers avoid rolling up their paper, for rolling slightly damages the fibers.

### *Distribution*

DISTRIBUTION methods differ throughout Japan. The papermaking house may transfer its paper to a cooperative or local wholesaler, which acts as a clearing house, selling to larger wholesalers. The manufacturing house may ship paper directly to the large, central wholesaler, from which retail stores and shops may buy; or the papermaker will work only from orders he receives from individuals or large paper dealers—this, particularly, if he is an accomplished and well-known paper craftsman. Naturally, the more middlemen involved, the higher the price of the paper when it reaches the consumer. Because of greatly increasing paper prices, some papermakers are forming cooperatives to regulate production costs and facilitate distribution to retailers and large wholesalers.

### *Properties and Definitions*

AS DEFINED in the first chapter, paper is an aqueous deposit of isolated vegetable fiber that has been broken down and felted upon a screen. Whether or not he realizes it in such terms, it is the problem of the Japanese hand papermaker to isolate the fibrous cellulose as thoroughly as possible from the noncellulosic matter, and to do so with as little chemical treatment and damage to the fibers as possible in order to produce flat-surfaced sheets of fairly uniform thickness. This approach, along with the fact that the fibers are kept long for strength, means that the paper he produces may lack somewhat in smoothness and whiteness, but it will excel in strength, toughness, and endurance. This describes most varieties of washi.

### *A Brief Chemistry of Paper*

ESSENTIALLY paper is an organic substance, the best of which is exceedingly high in cellulose content. Although differing greatly in physical structure, all plants contain tissue that, when properly processed, yields cellulose. Theoretically, then, any plant is a potential source of paper fiber; but the plants and parts of plants that naturally give a high yield of cellulose will be the most desirable for the manufacture of paper, since they require the least amount of processing to render them into suitable paper stock. No matter what fiber the papermaker uses, it is necessary for him to free the vegetable fiber from its nonfibrous (or noncellulose) constituents, reducing the raw material to a more or less pure form of cellulose. These nonfibrous elements are generally in the form of pectinous gums, lignins, resin, fatty and siliceous matter, aqueous extracts, and ash.

In the papermaking industry there are two important categories of cellulose-containing materials: (1) pecto-celluloses and (2) ligno-celluloses. Pecto-celluloses are mixtures of cellulose and colloidal pectin substances. Practically speaking, the pectin is separated from the cellulosic material by either alkali or acid treatment, after which the fiber is suitable as a papermaking material. Pecto-celluloses include straw, esparto, flax, and bamboo, materials that also contain some lignin.

The fibers most used in Japan, namely *kōzo*, *mitsumata*, and *gampi*, are ligno-celluloses; that is, celluloses bound by lignin. When the lignin is removed by treatment with an alkali solution at a high temperature, the remaining fibers consist of a fairly high percentage of pure cellulose. The middle and upper sections of the inner bark of good-quality *kōzo* average from 66 to 67 percent cellulose; *gampi*, 49 to 53 percent; and *mitsumata*, 56 percent. Compare these to cotton, which in its raw state contains 91 percent and is the purest natural form of cellulose; hemp, 77 percent cellulose; wood (pine), 57 percent cellulose; bamboo, 48 percent cellulose; and straw, 40–48 percent cellulose.

### *Physical Features*

CERTAIN physical features are used as parameters to describe a paper's capabilities and to aid in choosing the right paper for a particular purpose. Alongside a paper's chemical purity, these parameters indicate how a paper will endure time and the elements; more immediately, they help in knowing which papers will withstand the various stresses and strains to which one may wish to subject them.

Some of the various measures of paper strength are referred to by such terms as tensile strength, compressive strength, stiffness, folding strength, and elasticity. Roughly defined, tensile strength is a sheet's ability to be pulled lengthwise or breadthwise before rupturing. Compressive strength is, roughly, the force required to crush a sheet from an angle perpendicular to the sheet's plane. Stiffness refers to a sheet's resistance to bending. Folding strength is, of course, a sheet's resistance to folding. Elasticity of paper refers to a sheet's ability to recover its original dimensions to a greater or lesser degree after undergoing deformation. Indications of these various strengths are important factors in giving a picture of a paper's length of life, for they show the amount of wear and tear a sheet will tolerate.

Bulk is another feature of paper that should be considered carefully. Bulk is the thickness or volume of a pile of a certain number of sheets under pressure. The individual fibers of a bulky paper are considerably far apart, the sheets containing plenty of air space. Bulky papers are rough, light, and not very strong.

In addition, paper is spoken of in terms of insulating properties. Certain kinds of washi are such good insulators that they have long been used in Japan for sliding windows as well as for clothing. (The Japanese have long known of this paper characteristic, for they occasionally speak of a paper's "temperature.") Paper is also quite hygroscopic—it tends to adsorb moisture from the surrounding atmosphere. Naturally the amount of moisture contained in paper will affect many things, including its strength, flexibility, elasticity, and surface quality, as well as its rate of deterioration.

*Permanence and Durability*

TAPPI* defines permanence as "the degree to which paper resists chemical action that may result from impurities in the paper itself or from agents from the surrounding air"; and defines durability as "the degree to which a paper retains its original qualities under continual usage." Because these two terms are so closely linked, they are usually referred to jointly as permanence/durability.

Everyone involved with paper to any degree is concerned with paper permanence and durability. The Japanese have always been particularly concerned about the strength and durability of their handmade papers, and they have been justifiably proud of producing a great range of papers exceedingly tough and durable. Recently an American paper expert tested at random a sheet of Japanese writing paper (the exact type was not indicated). He was amazed to find that it has a burst of 1.7 points per pound per ream and a tear of 1.5 grams per pound per ream. This is exceptionally good. Much of the first paper ever produced in Japan, some quite thin and delicate looking, is today in good condition outside of museum environments, clearly bearing the ink and dyes inscribed upon it centuries ago.

The deterioration of paper has bases in both the chemical and physical structures of fiber and sheet. The purity of its fibrous elements and the lack of harmful chemical residues in the paper are just as important factors as the physical properties of the paper

* Technical Association of the Pulp and Paper Industry, Committee on Permanence and Durability

in determining permanence. The best papers, those that measure their lives in terms of centuries, are made of tough, thin, and fairly long fibers, such as the cotton and linen rag utilized in the West and the *kōzo* and *gampi* utilized in Japan. Papers of excellent, chemically unadulterated cellulose fiber of the kind described, manufactured to maintain maximum physical strength of the sheet, are obviously the papers capable of taking the roughest treatment and capable of enduring the longest.

*Chemical Purity*

As a rule of thumb, one can expect a paper to deteriorate almost in direct proportion to the amount and intensity of chemical treatment required to convert the raw fiber into cellulose. But the use of some chemicals in the paper industry, even in the small hand mills, can hardly be avoided. Certain chemicals help isolate the fibers from their noncellulosic constituents, and in so doing begin the physical as well as chemical structure alterations necessary in paper manufacture. The danger of using chemicals (aside from their obvious adverse effects on the environment in large-scale paper manufacturing) is in the chemical residues that remain in the sheets, which inevitably harm paper to a greater or lesser degree over a period of time because they upset its pH balance. This is particularly true of bleaching agents. A common consequence of the presence of chemical residues is acidity, paper's enemy. It attacks the cellulose, eventually weakening the fibers and discoloring the sheets.

The presence of alum as the Japanese sometimes use it in making certain "design" papers tends to throw the pH balance of the paper over to the acid side of the scale. Extended oxidation occurs, reducing greatly the paper's tensile strength. Contact of the fiber with iron is also a hazard in making paper and in dyeing it, since exposure of iron to air under certain conditions causes oxidation and subsequent discoloration. In Japan excessive alkalinity in the paper usually results from failure to rinse the fiber completely after alkaline processing treatment. Excessive alkaline residue is harmful unless it happens to offset paper acidity in the manufacturing process; this, it is believed, is what happened in European papermaking at various places and times, particularly the 1600s, when milk solutions and other forms of calcium carbonates were used to ret the fiber. The resulting papers are frequently still extant and in good condition, for it appears that a paper rather on the alkaline side of pH neutral, perhaps even up to 10.5 on the scale, has a built-in safeguard against "natural" acidification over time.

In bleaching the fibers, Japanese traditions have long employed nature to help with the whitening process. By using running water of streams, sun, and snow, papermakers have successfully bleached fiber white without benefit of chemicals. As a result, it is often observed that *kōzo* paper whose fibers have been bleached "naturally" turns whiter with age; whereas that bleached with chemicals goes the way of most paper, turning yellow. Bleaching powder is often used today in excess or it is not rinsed thoroughly from the fiber. In either case, the result is paper containing hypochlorite residues, which inevitably weaken paper structure. It is a pity that these time-proven whitening methods are not practiced more often, even in Japan.

As an example of what may happen when paper is overtreated with harsh chemicals, a friend, an American living in Tokyo, told me of one rainy evening when she and her husband arrived at their calligraphy class. As they unrolled their calligraphy paper, their Japanese master gave a little cry of astonishment. Coming over to them he picked up the sheets, looked carefully at them and announced, "Your paper has caught a cold!" Their practice sheets, he explained, were made with such excessive use of chemicals that when they were exposed to the humid air that night they adsorbed some moisture, buckled, fell limp, and in general "got sick."

In Japanese paper hand-manufacture, the excessive use of chemicals, especially such harsh ones as bleaching agents and alkalis, is unfortunately on the increase. Generally, the effect of chemical residues is felt over periods of time, often long after the paper's manufacture, and is of no little importance to the artist and printer alike; it determines not only the permanence and strength of the paper but the quality and endurance of applied ink and color as well.

Substances called loads are frequently added to paper to fill the pores for a smoother surface and to add weight to the finished sheets. Loads or fillers such as china clay and other powdered minerals are commonly added to the wet stock of certain types of washi. Loads of this type will substantially decrease the paper's strength and elasticity, and therefore its durability, especially if used to excess. However, this is not a serious problem in Japan, as far as I know. Few papers are loaded, and most of these not to excess; this technique is limited to special papers in which a smooth, slick, opaque surface is required or papers backing old works of art, where the filler helps draw harmful moisture out and away from the old paper.

### *Factors Determining Durability*

Two papers of the same chemical pureness may differ significantly in physical strength. The length of the fibers, their physical condition after beating, and the manner in which sheets are formed all determine much of the paper strength.

Beating alters the structure of fiber physically as well as chemically. It is an important step in paper manufacture, for this is when the fibers are modified and hydrated. Here the relationship of cellulose to water is vital, affecting the strength, flexibility, and shrinkage of fibers as well as their capabilities of being formed into sheets. The length of time that fibers are beaten determines their degree of hydration; up to a certain point, the longer beaten, the more hydrated they become, meaning increased moisture retention and slightly increased cohesiveness. Naturally, fibers must be beaten long enough so that they separate into single strands and result in a wet pulp that can readily be formed into even sheets.

The manner in which beating is carried out is important to fine papermaking not only because it establishes the initial cellulose-water relationship within the fibers but also because it determines fiber length and quality of maceration. When fibers are beaten by hand using bats or mallets, as they were everywhere in Japan until recent generations, they are not so much cut up or disintegrated as separated, bruised, and teased out. Such beating produces paper of extraordinary strength. The hand labor involved, however, makes hand-beating prohibitively expensive, and some papermakers regret reverting to machine beaters but have little choice if they are to make any profit.

Mechanical wooden stampers, quite similar to those used in European paper manufacture in the middle ages, are considered a good substitute for hand-beating by paper experts, but few papermakers employ them in Japan. Use of the Hollander-type beater, however, which is commonly seen in small hand mills throughout Japan today, has been shown to impair the durability of paper somewhat, particularly its fold and tear resistance. While not terribly harmful to the fibers, the Hollander-type machine nevertheless does tend to shorten them, cutting them across the grain as well as splitting them lengthwise, and may not promote cohesiveness as well as does hand-beating or the use of mechanical beater-stampers.

Next to beating, the pressing and drying of wet sheets are important steps in the formation of the physical structure of the sheet, largely affecting its bonding properties and the amount of moisture it contains. Normally paper contains a certain percentage of water as part of its physical makeup. Without a certain amount of moisture paper would be too brittle to use. Therefore the amount of moisture present influences paper's strength, elasticity, and durability.

Pressing is performed by a gradual increase in pressure and weight. Because the sheets lie directly one on top of another in washi manufacture, pressing must be executed with great caution and discretion lest the sheets adhere or dry too quickly or unevenly.

Whereas Western handmade papers dry slowly and naturally indoors, Japanese papers are brushed onto boards and carried into the sun and wind, where they dry quickly and crisply. Because this process is so much at the mercy of the weather, papermakers in recent years have taken to using indoor dryers—heated iron sheets—onto which the paper is brushed. Paper dried in such an artificial manner loses moisture too quickly. When wetted and used in various types of printing, this paper then tends to stretch out of shape, making good registration difficult.

Papers also deteriorate by the action of fungus and insects that feed on starch sugars present, notably those papers either made of certain kinds of cellulose, sized with gelatin, or containing nitrogenous material. Fungus growth on paper is especially prevalent in hot, humid climates and weather—such as Japanese summers—but this kind of growth feeds largely on gelatin sizing, which is not commonly applied to washi. The presence of starch can produce similar results. Paper made from straw, because of its type of cellulose, is more likely to be attacked by organisms than that made from bast fibers.

NOTES

half-title: Tokutarō Hanada and Kiyofusa Narita, *Kami-suki uta*, 2 vols. (Tokyo: Seishi Kinenkan, 1951). (group trans.)

1. Sōetsu Yanagi, "Washi no bi," *Ginka* 9 (Spring, 1972): 32–36. (trans. by Louise Picon Shimizu)

2. Bunshō Jugaku, *Paper-Making by Hand in Japan* (Tokyo: Meiji Shobō, 1959).

3. Yanagi, *op. cit.* (trans. by Louise Picon Shimizu)

*These hands making paper—*
*Repeating and repeating,*
*How many times*
*Over and over again*
*As the winter day dawns.*

# The People Who Make Washi

THERE ARE FEW CRAFTS in Japan that involve so much plain, hard labor as papermaking, yet also have elements of the genteel and artistic. Papermaking has always been monotonous, back-breaking work; the pressures are many and the gains few—a papermaker's efforts are rewarded in dividends of few joys, fewer holidays, and little money.

Much of the work is heavy—harvesting, toting around materials and large boards, the steaming, and even the forming of the sheet on the mold—often a woman has to balance and toss around many pounds of solution on the screen at one time, over and over again throughout the day. Japanese papermakers have had to be hard workers just to survive in their craft, and have a reputation as being such. A friend of mine living near Mt. Fuji once told me a proverb his mother used to recite to him as he approached the marriageable age: "If you're looking for a patient and hard-working woman, go to Kajima on the Fuji River where people make paper."

The Japanese have a strong philosophy of work. Since ancient times the papermakers' lot has been so hard and their living conditions so harsh that it seems they have had little choice but to resign themselves to it all. For most, papermaking was a livelihood they inherited, not one they chose. The Japanese response to their situation showed greatness—they composed their lives into small works of art. By sincerity, faith in nature, and a kind of reverence for his craft, the papermaker transcended the meanness of his condition, and somehow his small existence took on greater dimensions. He took pride in his paper—the best of it would be the treasured possessions of the nobility—and he felt a certain beauty when he contrasted his humble life from which the paper grew to the exalted, luxurious lives in the courts.

> I will choose a papermaking girl for a wife.
> Her complexion is as white as the *hōsho* paper she makes.
> All the nobility, even our lord and the shogun, use the
> *hōsho* paper of Goka.
> Although others long for the festival at Bon season, I look
> forward to our year-end holiday.

The courtesans of Goka make do with hurried meals of
rice and tea.
Their skin is white and they are beautiful, but also limp
and cold.
Girls of Goka live on rice balls and make love tenderly.
We have the heaven-sent task of making paper from
generation to generation.[1]

The papermaker is not required to be a creative artist, nor a craftsman of great aesthetic vision. He is not, like the self-conscious artist, driven by the idea of creating Something of Beauty. Tradition is for craft what inspiration is for fine art—the foundation of excellence. It is faith in tradition and intimacy with nature that allow the papermaker to create beautiful sheets. As Sōetsu Yanagi said, "tradition bears the burden"; the papermaker works "eventlessly." One sheet looks quite like the next—the papermaker does not carve his personality into every sheet, nor would doing so glorify his paper or improve it. The beauty of washi is an anonymous beauty. "There is no ego in paper," Yanagi wrote.

My teacher, Seikichirō Gotō, once told me that even a bad person, or an unskilled craftsman, can make good paper—if he makes it with sincerity and honesty. For it seems as if everything is subordinate to the work itself, while in the work something of a man's more universal nature finds its true and natural expression. Yanagi's formula: "Natural material, natural process, and an accepting heart."[2] It is almost invariably true that the further the village from the mainstream of modern life, the more beautiful is the paper made there—for there the people still live close to the land and there tradition has not been lost or sullied.

Papermaking is a cooperative effort, usually involving whole communities. The lone papermaker is a rarity, for the work is heavy, many-faceted, and sporadically urgent. Because waste of time, effort, and materials means failure, close cooperation among workers who take their roles and responsibilities seriously has always been vital to the hand paper profession. Like guild papermakers of Renaissance Europe, papermaking is a family craft passed on from father to son—although in Japan just as frequently from mother to daughter. Often it is even necessary for the young children to help out with the work, and most begin learning the craft seriously in their early adolescence.

Although papermaking in Japan is a cooperative effort carried on generation after generation by anonymous craftsmen, in some cases one individual has emerged out of a body of paper craftsmen and stood out from them due to the force of his personality and the superior quality of his paper. They have become heads of famous papermaking houses—men such as Naruko of Ōmi; Endō of Shiroishi; Shimizu of Misumi; Abe of Matsue; Furuta of Mino; and the late Iwano of Ōtaki.

In this chapter some of both these types will be described—the unknowns and the "greats." It will deal with the people in the town of Kurodani, Kyoto Prefecture; and with three famous papermakers of different types—Eishirō Abe, Ichibei Iwano, and my teacher, Seikichirō Gotō. It is a cross-section of the people who make paper, one that I think will give impressions of their work, lives, and thoughts.

Kurodani is not so much a "typical" papermaking village, if a typical one exists, as it is an "ideal" papermaking village; it is anyone's expectation of what an idyllic Japanese country hamlet should be. The papermakers of Kurodani are warm, relaxed, ingenuous, somewhat shy. These people, and the sounds, smells, and colors of the village are unforgettable. Everyday events are the treasures of my stay there: gay children and gossiping women pausing as they pass on the lane next to the village river, carting pulp home after beating; a woman in a tiny station wagon chugging up the lane selling fish and *tōfu*; walks among the sun-drenched houses, listening to the distinctive sloshing of water over the paper screen. Bits of paper fiber on the windows, on women's aprons, in their black hair; the sensuous vegetable odor of bark permeating musty rooms; workers running out to carry in their drying boards as a summer shower suddenly begins to fall. Kurodani is like one rather large and congenial family.

The three individual papermakers are another matter entirely. All are first-rate craftsmen, enjoying some degree of fame and respect in their field and among other artists. Iwano, who died in November of 1976, was a man of the earth, clear eyed and clear headed. His masculine paper was stout and strong hearted, simple, tough, resilient, and will last a long, long time. Abe is primarily a man of taste. Like Iwano, he was born of a long line of farmer-papermakers, but his papers, home, and workshop reflect a beautifully mellowed and refined elegance found nowhere else in the paper world. Gotō is first and foremost an artist-craftsman. He began making paper rather late in life as a side vocation, becoming increasingly involved in it over the years. He is probably the only Japanese papermaker who is also recognized as a top print-maker and dyer as well. These men are beyond attempts at comparison—all are superior craftsmen making paper with great sincerity and integrity, making it in a way that suits their natural bents and talents.

### *Kurodani*

THE village of Kurodani lies along the north-bound road between Kyoto and the Japan Sea on the cold, northern edge of the old province of Tamba. This is one of the most beautiful and prosperous papermaking villages in Japan. The name means "Black Valley," which, the villagers state, is because the sun rises late in the morning and sets early over the mountains that steeply surround the village. Soot-black steam locomotives until recently roared through the valley. Age is inscribed everywhere in the gray wooden houses, their thatched roofs encrusted with luxurious green moss and dotted with wildflowers. Snow falls early in winter here, and rains beat down heavily the year through; but the feeling of Kurodani and her people is of great vividness and warmth. After school in spring the children play in the thick bamboo groves on the ridge behind their houses and dig for bamboo shoots to bring home for dinner. Wild boars nibble at the *kōzo* buds and potatoes in the family gardens; and in autumn a villager drying his paper on the slopes may suddenly come across a bear stealing persimmons from the hilltop trees.

The history of Kurodani is not well known, but the village is certainly very old. There is an established theory that the original inhabitants were members of the famous Heike clan who, after their defeat in the Battle of Dan-no-ura (1185), fled and scattered into the Tamba hills. At that time there were about sixteen houses in Kurodani; now there are fifty-four, with nearly all the inhabitants claiming descent from the Heike founders. How papermaking technology first came here is lost to history, but it is a likely adoption, for the steep slopes discourage the planting of rice, and papermaking fiber shrubs once grew wild and plentifully. Kurodani sits along an ancient road to Kyoto, so that even in early times it was convenient to trade. Records show that the Tamba area was sending paper, along with cloth and plant fibers, to Kyoto as early as the Heian period. This area is known to have been particularly prolific in its paper production around the 1600s and after. Papermaking is now a full-time, year-round way of life in Kurodani, with about eighty people making paper—about thirty-six of the fifty-four houses—and another forty people doing paper-related work.

### *The Cooperative System*

MOST of the town's activities center around the *washi kumiai*, the papermakers' cooperative union. The union was founded just over one hundred years ago, during the early Meiji period, probably as a defensive measure against the new competition of machine-manufactured paper. Today it is an important and highly effective arm

of the people, acting as an agent in supplying raw materials and tools from other parts of the country to the papermakers and in selling and distributing their products. The cooperative also provides and keeps up expensive large-scale equipment in the communal workshop and steam drying rooms that the papermakers could not afford individually; it invests a portion of the workers' profits in the district's agricultural cooperative; and finally it subsidizes those papermakers who are in need of money. Kurodani's *washi kumiai* is one of the most advanced and smoothly run in Japan.

Heading the union is the union chief, who is elected by the community of papermakers at large for a three-year period. Hajime Nakamura is presently its head and has been since 1964; he is partly responsible for guiding Kurodani out of one of her toughest periods in the wake of World War II. Directly under Nakamura are three committee officers, who each serve a three-year term as well. They act as consultants and advisors to the villagers for all imaginable kinds of problems, from the shortage of materials to a papermaker's personal dilemmas. Aside from these officers, there are four secondary workers in the cooperative office, one in the communal workshop, and eight in the craft studio. All are full-time employees of the cooperative and, with the exception of the man who manages the beating of fiber in the communal workshop, all are Kurodani residents. The cooperative holds two formal meetings a year to report on its activities and other "important gossip." Aside from these, lesser meetings are often held, especially just before April election time.

Orders for paper come to the union from all parts of Japan; usually no paper is made until an order is received for it. Along with the steady orders, the union might receive a special order—such as when a typhoon hit the famed Katsura Imperial Villa some years ago and *fusuma* paper was requested for repair work. Such orders are then divided up and delegated out by the union. There are no special rules or regulations governing this work of delegating; but the cooperative knows the talents, particular skills, and equipment of all the families, and this information is the basis for the decision. As a result, people are occasionally asked to do something they might rather not do, but Nakamura says that so far this has presented no problem. The papermakers accept and respect the union's requests. Where special knowledge or subsidization is required to help a papermaker do the work delegated to him, the three advisors come into play. And the union often rents out special equipment, such as frames and screens, to help a papermaker make an unusual kind or size of paper.

A cost analysis in which all the factors determining price are estimated—raw materials, labor, the manufacture or purchase of special equipment, etc.—is drawn up by the union and submitted to the papermakers for approval. This cost analysis is calculated from an index drawn up and established by the union, which can measure their labor more accurately than the papermakers themselves. Usually some adjustment of this estimate is made between the papermakers and the union. Then the provisional estimate is submitted to the orderer, who may approve the price, reject it outright, or bargain further. Not infrequently the actual cost of making the ordered paper is higher than the agreed price; so if this becomes apparent as the paper is finished or nearly so, the union approaches the orderer again until a settlement satisfactory to all is reached.

When an order has finally been settled upon, the union begins by selling the papermaker raw materials. Some families in Kurodani have inherited *kōzo* fields of their own, but most have no such supply; and even if they have their own *kōzo*, it is usually only enough for about three months of papermaking. Moreover, the harvesting and processing of the bark is laborious and time consuming. The railway and highway occupy the best valley land adjacent to the village, largely forcing *kōzo* cultivation into the hills and away from easy access. Therefore about 95 percent of the *kōzo* used in Kurodani is bought and brought in from Shikoku, Okayama, Wakayama, Ibaraki, Nagano, Kyushu, and other parts of Tamba, and even Korea. Fibers such as *mitsumata* and *gampi* are also trucked in, but on a much smaller scale.

Those who own *kōzo* fields usually also have the equipment for steaming it. This is often nothing more than a huge pot and an equally huge wooden barrel to cover it, holding the packed-in *kōzo* shoots upright as they cook. Those who do not have this equipment either rent a neighbor's pot or pay him some agreed-upon fee to do their steaming. Most of Kurodani, however, depends upon imported *kōzo*, which has already been steamed off the shoots, so this type of equipment is used only by a few. Kurodani's home-grown *kōzo* is some of the finest grown in Japan. The papermakers grow and prepare their own mucilage locally. The bark of a tree known colloquially as *sana* comes from the forests on the surrounding hills; and the *tororo-aoi* grows in the home gardens between the cabbages and the radishes.

All the villagers make free use of the two river branches that flank the town. The water from this river is of prime importance to the town's papermaking activities, although Nakamura says that today it is only half as wide as it once was. Upstream the papermakers place their *kōzo* fiber under large stones in the water where the river is dammed up. The next day they wade into the stream in rubber boots, treading on the fiber over smooth, flat stones; this loosens the flaky black bark. One old woman told me that during the war her daughter had no boots, so in winter she trampled barefoot over the *kōzo* in the ice-encrusted stream—the same method reputedly used in early days. Downstream, on the north side of town, all the chemical dregs and residual dyes are dumped into the river from the communal workshop.

The communal workshop is a fairly large structure housing equipment for boiling, washing, *chiri*-picking, bleaching, and beating the fiber. It also contains some simple facilities for dyeing paper, especially for *shibori-zome*, tie-dyeing. The workshop is for all members of the Kurodani union. A papermaker who wants to use some particular facility makes an application usually at least a day ahead of time, but sometimes on the same day he wishes to use it. Because the papermaking process is cyclical and good timing is so important in maintaining the vitality of the fiber, the workers make efforts to keep the workshop schedule running smoothly.

Equipment for boiling the papermaker's white bark (*shirokawa*) in an alkali and then washing it lies in one section of the workshop. This is usually done by the workers themselves, but they may pay the full-time workshop employee 200 yen (60 cents) to boil twenty-five kilograms of dry bark.* This mainly pays for the fuel, since the work involved is largely a matter of stoking the fire from time to time over a period of several hours. Sometimes bleaching powder is added. Sinks for washing out the alkali and bleach are nearby. Often a papermaker will spend most of an afternoon at the sink washing a fiber that has been boiled without bleach over and over again for special high-class white papers. This laborious rinsing has the same effect as, and is much preferable to, bleaching with chemicals.

In front of the workshop and facing the river is an area sectioned off for *chiri-tori*, the picking of impurities out of the boiled, softened fiber. In Kurodani this is a roofed-over area with a sluiceway dammed up before it. Kneeling on cushions placed on the cement, the women lean over great open-weave baskets of fiber that float in the water and pick away at bits of black bark with their fingers or pins. If a family has a grandmother, this task is generally delegated to her, as it requires patient and persevering attention rather than great physical exertion.

The beating machines take up most of the indoor workshop area. Unfortunately hand-beating no longer is practiced in Kurodani, just as this technique has disappeared almost completely throughout Japan. Two varieties of mechanical beaters are used here; one, a massive wooden beater with several hammers for the preliminary beating; and the other, a metal and concrete Hollander-type beater, for finer work. The union employs a full-time beating expert from a nearby town to work these machines. A papermaker pays 400 yen (1.20 dollars) for him to beat twenty-five kilograms of dry white bark, which includes a preliminary and then finer beating. Even using these mechanical beaters, it takes about four or five days to beat six *kan* (about twenty-

* Prices and monetary rates of exchange here are based on 1973 figures unless otherwise stated.

three kilograms) of dry bark—which will fill a box of about 75 × 60 × 60 centimeters (2.5 × 2 × 2 feet) with wet bast. Every day some women will be pulling the wet, beaten bast from the Hollander vats and toting the heavy tubs over to the open cement tanks for further *chiri*-picking. After thoroughly inspecting the floating bast for dark bits of bark, the women drain it of excess water and cart it home in small two-wheeled pull wagons.

All the houses in Kurodani have a workshop and basic scooping equipment, such as a vat, a comb, and a variety of new and old screens and frames. If a customer orders through the cooperative some special paper such as one with a particular watermark, the union commissions someone skilled at making watermark designs to prepare a screen. This screen, or any other special equipment, is rented out to the papermaker. Most screens and molds used in Kurodani are made in Kōchi in Shikoku. They are not cheap—100–120 dollars for a medium-sized screen and frame combined. Sometimes the papermaker will not even get his money's worth out of one. I observed that one woman, who was using a Kōchi mold, poured some additional solution into one side of the frame after each scooping. She explained that the frame was not well balanced. The woodwork of the frame was well constructed, she said, but the carpenter was not a papermaker, and so had not realized the delicate balance so essential for his product to function well.

Water is basic to papermaking, and Kurodani is fortunate to have its sources of good water. Since the house and studio often flow one into the other, the papermaking vat usually shares a faucet with the washing machine. No one lives very far from one of the two river branches. Houses that lie against the hills also funnel water into a backyard tub from one of the many mountain spring-fed streams by way of U-shaped split bamboo ducts. This water is cleaner than the river water, and so is very good for soaking and rinsing the fiber.

Every house that scoops has an area in town where the papermaker sets up his drying boards—it may be his front yard or it may be an area high on the ridge overlooking town. There is inevitably a small shack nearby, where the boards are stored and where the papermaker can brush the damp sheets onto the boards away from direct sunlight. In summer, wild monkeys often sleep in the shacks highest up on the hill. No one in town knows where his boards are from; they have been handed down through the family. Generally they are of pine, with the grain worn into soft relief through years of use. Paper dried on such boards comes away embossed with this lovely wood pattern, which Kurodani papermakers appreciate and use to good effect. For fine, delicate papers that require a smooth surface, boards are planed and sanded.

Because Kurodani's sunlight hours are short and because it rains so much of the year, the cooperative maintains an indoor community drying house. The dryers are sheet iron, two such sheets set at an angle to form a standing triangular shape, and are heated from below by either wood or oil. These iron sheets dry the paper quickly, but the crispness characteristic of sun-dried paper is lost—and of course there is no wood-grain pattern. Generally the papermakers do the drying themselves, paying a fee that mainly goes toward the fuel. In winter and during the summer rainy season, the drying room is in constant use. A papermaker, although at the mercy of the weather, tries to reserve a dryer ahead of time by marking his name on a calendar in the drying room.

The cooperative union pays Kurodani workers to do some piecework connected with the finishing of certain papers. Kurodani produces much *momigami* (see p. 205), a very strong paper made stronger and somewhat water-resistant by coating it with a vegetable starch solution and then wrinkling it between the hands. Some residents, too old for the exertion of most papermaking tasks, do this in their homes and thereby make a simple living. The union supplies them with the paper made by a Kurodani papermaker, sells them the powdered starch, pays them 17 yen (5 cents) for each sheet they treat and process, then markets the *momigami* for 200 yen (60 cents). Full-time workers doing only *momigami* processing can make a very modest monthly income.

Another example of piecework is *shibori-zome*, tie-dyeing. Housewives can make extra money by tying knots in the *momigami*, sitting before television in the evening or in good weather sitting on the porch gossiping with their friends. The union supplies them with the paper, and chemical dyes are available in the communal workshop. When the union receives the dyed sheets, it pays the workers according to the number of knots.

In the handcraft workshop behind the union office, eight full-time union employees turn out wallets, purses, vests, kimono wrappings, book covers, and other items out of the patterned *momigami*. Most of the paper used here has been sent to Kyoto for stencil dyeing and returns to be made into various washi items in this workshop. Occasionally some women do *itajime-zome* (fold and dip dyeing) in a back room. Low tables line the straw mat floors; here the women sit, cutting out purse forms, ironing, gluing, and sewing the paper into purses, pocketbooks, cushions, and even paper carpets. The one man employed does the more rigorous work of pressing and binding notebooks fashioned of Kurodani paper.

The papermakers sell their finished products to the cooperative. Eight percent of the total amount they would receive is withheld by the union and deposited in the district agricultural cooperative bank. Here the money sits for one year, collecting interest compounded annually at 5.6 percent. At the year's end, the union presents the papermaker with the 8 percent plus its accumulated interest. The papermakers pay no membership dues to the union; instead, the profit margin made by the union, acting as an agent in selling the paper, is considered the dues. In the event that the profit is unusually large and too great to be considered dues, this amount is also deposited in the district agricultural bank, where it collects interest. This capital is treated as a fund from which the union may withdraw portions at any time it deems necessary, usually to help subsidize a papermaker in need of money.

Thoreau's "that government governs best which governs least" seems to describe the Kurodani cooperative union. On the surface anyway, the union appears to provide maximum benefits for its members with a minimum of restrictions. Kurodani papermakers are as hard working and conscientious as any, but no one is under apparent strain or hardship. All the villagers are extremely cheerful and healthy looking. They come and go as they please, maintaining hours of their own choosing. One man can always be heard scooping paper late at night, until eleven o'clock or midnight, when everyone else has retired by nine. "Oh, I'm not all that industrious," he insisted. "It's because I like to sleep late in the morning and waste time." Where the union seems wisest is in its ability to organize, delegate, set deadlines, and in general control the flow of capital, yet also respect the papermaker's need to regulate his own life and work schedule.

In reality the union functions as much more than the town's paper broker. It is the spiritual core of a village that is in many ways a kind of large family unit. The feelings of the village people toward the union are an outgrowth of old feudalistic Japan and its collective orientations, made even stronger in Kurodani by the fact that everyone here with few exceptions claims descent from the same founding clan. In the succeeding centuries of intermarriage, a deep sense of community and homogeneity has developed quite naturally among the villagers. At the center, the union acts as the stable and directing force of the community; its chief inherits the role of village patriarch along with that of the town's business and administrative leader.

*The Horii Family*

ALL three generations of the Horii family were born and raised in Kurodani; their lives have been intimately related to handmade paper. Indeed, papermaking is the only way of life they know. Even the raising of silkworms, which had long been important as a means of livelihood in the weaving country of Tamba, eventually died out here as papermaking became a year-round activity. The Horiis are to a large extent typical of nearly all papermaking families in Japan. Their life-style has changed considerably since the early 1960s, reflecting the modernization that has irrevocably changed Japan and, in so doing, also forecast the death of the way of life that for centuries has made the production of washi possible.

The Horii family lives in a big old house resting against a bamboo-covered hill on the east side of the village. The land it lies on used to belong to Hanjirō Mizuguchi, who founded the Kurodani cooperative union at the beginning of the Meiji era. Mizuguchi's only son moved to Kyushu and died there. At this time a young girl by the name of Asae was living next door. She married one of the Horii sons, and they moved to the nearby port city of Maizuru for a while. After a time, Asae's father became lonesome for her. He bought the Mizuguchi land and built a big, rambling house on it, and thereby induced the couple to return to their home town. In later years the son-in-law himself became head of the cooperative and was responsible for helping pull Kurodani papermakers through some of their toughest years in the aftermath of World War II. In 1973 Grandpa Horii was sixty-two, and Asae, Grandma, sixty.

Their son and present head of the household is Shun'ichi Horii, forty. Every day he and his father commute to Maizuru, where they work at a dockside lumberyard, unloading, cutting, and measuring logs off Russian, Canadian, American, and Southeast Asian ships. Although papermaking is their inherited craft, in the past century it has never been a profitable profession; the family made the decision in 1959 to seek work outside the village, as many others were already doing. Gradually papermaking became the work of the women and slid into a secondary position of importance as a source of family income.

Chisako Horii is thirty-six. Traditionally women have always played a very important role in the family papermaking; in Kurodani today women like Chisako are trying to carry most of the burden alone. A female neighbor of hers commented wryly that if any man moved into Kurodani it would be an easy life for him. And the pace is very tough on Chisako. She is wife and mother as well as full-time papermaker. She is the first up in the morning and the last to retire at night, with very little rest between.

There is an old poem written by some sympathetic husband of such a woman:

> My wife has become used to
> Always working so hard making paper;
> But once in a while she lets slip
> How really tired she feels.[3]

The Horiis have three children, all girls. Chieko is thirteen and attends junior high school; Mieko is ten and in grammar school; Sachiko, six, goes to kindergarten. The techniques of making paper have long been handed down in paper towns from mother to daughter along with all the other feminine and domestic arts. Normally girls of Chieko's and often even Mieko's age would have already begun to learn scooping as well as other papermaking skills. But the Horii girls have no interest in papermaking, have no desire at all to learn a craft that will guarantee for them a life of extreme hard work and sacrifice. Nor do their parents wish such a life upon them. Chieko, who studies English at school, wants to be an English-language interpreter. So in these children and their young neighbors the chain has already been broken.

The day starts at 6:00 A.M. for Chisako Horii. She gets up, dresses, and begins making box lunches for the children and her husband. Grandma Horii rises, dresses, and prepares tea to offer at the family altar, which stands in a side room. Sometimes Chisako joins her in praying before the ancestral shrine. Only the women keep up these religious duties. Afterward the two women make breakfast, which usually consists of soup with beancurd, raw eggs over rice, seaweed, pickles, fish paste, vegetables, tea, and occasionally some fresh fish—a typical rural breakfast. On chilly mornings the family sits on the tatami mats with their legs under the *kotatsu*, a low, blanket-covered table with a heating unit below it. If it is really cold, the small oil heater is brought out and turned on. At breakfast, the television is the center of attention. Except for the children, it is one of the family's few opportunities to see it. Sometime around 7:30 a message may suddenly gurgle out of the public address system from the union center, announcing a meeting or other important event.

The port of Maizuru, where the men work, lies about one-half hour's travel time north of Kurodani. The family does not own a car, so at 7:10 Grandpa starts to work on his motorcycle. His son follows at 7:30 on his—through rain, sleet, snow, and the bitter cold of winter. At 7:40 Mrs. Horii sends off the two youngest children, who catch the local bus to their school down the highway. At 8:00 Chieko leaves for school on her own bicycle. Grandma and Mrs. Horii wash the breakfast dishes, clean the house,

and do the laundry. By nine o'clock they are ready to begin the day's papermaking.

Mrs. Horii works about ten hours a day every day of the week making paper. Beginning at nine in the morning, she works until nine at night, with thirty minutes for lunch and two hours reserved for preparing, serving, and cleaning up after dinner. Grandma Horii helps her with the papermaking and the domestic chores she cannot get to. Most of the year Mrs. Horii fills orders for a type of soft white washi postcard with a wood-grain texture, called *mokume-iri*, which is her specialty. With the help of her mother-in-law, she can make about nine thousand postcards a month. The union pays her 7.5 yen (2 cents) per postcard. Expenses amount to about 20,000 yen per 10,000 postcards; she can usually clear around 50,000 yen (150 dollars) monthly. This is with Grandma and occasionally her husband helping; a comparable city job salary would pay her much more for her labor.

Mrs. Horii says that sometimes she feels like stopping her papermaking and just concentrating on her housework—but she has been brought up making paper and she is used to being constantly busy. "If I had so much spare time, I wouldn't know what to do with it!" she laughs. She says that women mainly do the scooping in papermaking, because it suits them. "Women persevere more. And they seem to have an intuitive 'feel' for handling the frame in the water. It takes much more than practice. Only a few men in town are good at scooping. They don't seem to take to it so naturally." Perhaps it is dipping the hands in and out of the cold water in winter that the men cannot endure; or constantly fishing insects out of the solution and picking them out of the paper on summer nights when the indoor light attracts them. Or perhaps women just have a gentler, more rhythmical touch. But she does not mind what others find so annoying, because she loves paper and papermaking. "If I didn't love it I'd probably hate it!" She just wishes she did not have to work at night too.

"Today's young people can't stand the hard labor involved. They haven't the endurance and perseverance you need for good papermaking. In a way I can hardly blame them. You should see how my hands crack with dryness and form sores. Sometimes the girls help me a little with the work, but they can't scoop and I doubt they'll ever learn. I can understand why they don't want to. If the income was higher and the

working conditions improved, there might be the possibility they'd learn. But I think the chances that would happen are slim."

Mrs. Horii feels there is a limit to how the working conditions can be improved without ruining the quality of the paper. "The process of making paper isn't *that* difficult. It could be mechanized—it *has* been mechanized," she said. "But each time you mechanize even one small process, the paper begins to lose something." Next to her paper vat is a little electric propeller for blending the vat solution each time it is replenished with raw materials. But she never uses it; it cuts the fibers too much. Instead she beats it with a large wooden comb. She laughs. "Maybe some day when I'm old and feeble I'll have no choice but to use it!"

Grandma Horii helps her daughter-in-law as much as she can, but since she is getting old and has a kidney ailment, she stopped scooping in 1972. Now she helps by scraping off black bark, picking away *chiri*, and aiding in the drying of paper. She has been making washi since she was fourteen—more than forty-five years a papermaker. She would stop now, but her daughter-in-law needs her help. And she still has one unmarried son—both she and her husband feel they must work hard to save something up for his marriage. Grandma Horii says she always was and still is fond of making paper—"If I didn't like it I surely would have quit long ago!"

Shopping is not very good in Kurodani. The only store in town is a small dry goods market in the cooperative office. Since the women cannot get out to shop, the shops sometimes come to them. A woman drives into town several times a week in a tiny station wagon loaded down with various kinds of fresh beancurd, fresh and dried fish, *konnyaku*, and other perishable foods for which the women would otherwise have to take the bus into the towns of Ayabe or Maizuru to obtain. The little white car drives

up and down Kurodani's one main road playing a jingly tune, and the women pause from their work to run out and stock up for the next couple of days. The vegetables they do not grow themselves they buy from a local farmer who carts them in by tractor. From time to time a local store sets up a sale of certain goods in Kurodani's special display house. Originally this lovely building was intended for entertaining honored guests and for showing papers, and between the bedding and kitchen equipment sales it is still used for this. Sometimes the women take time off to go shopping in Ayabe on Sundays, but it is never as often as they would like. The last time Mrs. Horii got to Kyoto, one and a half hours away by bus, was a year ago.

Carting their bast home in pull wagons, the women always find time for a short chat with a neighbor, pausing in the shadow of a house, where *kōzo* bark and strung fruits and vegetables hang drying together under the eaves. Because the sun sets early behind the steep hills, the workers must take in the heavy drying boards between four and five o'clock. And if there is a sudden downpour or snowfall, the women dash out to save the paper before the wet can damage it. Every so often a woman arrives in town from Ayabe to teach flower arrangement, so those women who can afford to do it drop their work for a couple of hours. Sometimes the wives get together to learn cosmetics or cooking; for Mrs. Horii this happens maybe once a year, although she enjoys it very much.

In the afternoon the children come home from school. They play in the bamboo groves behind their home or watch television and read comic books. Sometimes they help with some simple papermaking tasks, but this is rare. Aside from homework, their responsibilities are few. About 8:30 every night the children in town gather, walking down the main street clapping together wooden blocks to remind all the villagers to turn off their gas and put out their fires. Once every summer their parents take them to the beach. Although it is less than an hour away, the family cannot afford more than one short holiday in terms of both time and money. This, the parents feel, is perhaps the hardest sacrifice of all in being a papermaker—they want to do so much for their children but can do so little. And it is the same for all the families here, the Horiis say.

In the early evening Mr. Horii and his father arrive home on their motorcycles. This is their time to relax and play with the children. In spring, when they return while it is still light, they go to tend the fields. The Horiis own their own hill in town, about two and one-half acres of cultivated land where they grow rice and vegetables. In addition, they are lucky to have an acre of *kōzo* field on which Grandpa Horii grows enough trees to supply the family with a few months' worth of bark every year. On Sundays Grandpa takes a cart into the woods to chop logs and gather dry bamboo as fuel for heating the family bath.

Mr. Horii helps his wife with the papermaking when he can, mostly on Sundays and holidays. He knows every aspect of the craft, but, like most men, he cannot scoop well, so he leaves this to his wife. Along with the other husbands, he assumes community responsibilities, such as drilling in the fire brigade. And he does his part as an officer of the town's Self-Governing Committee, which largely serves to coordinate festivals such as Obon and the boys' summer sumo tournament. He fulfills his duty to his ancestors and gods by visiting the Shinto shrine on behalf of the town. Early every morning two people selected by the village committee go to the shrine to give prayers and offerings for the protection and prosperity of the town and its papermaking.

Everyone in Kurodani worries about the future of washi. Grandma Horii worries about it, and so does Mr. Nakamura at the cooperative union. Mrs. Horii has no faith anymore that it will endure. Her children are not learning the craft, nor are the neighbors' children. Mrs. Ishizumi next door, who is in her seventies, told me with some ironic sadness how years ago her daughters had promised her that when they grew up they would work hard like their parents, learn papermaking, and pass on the techniques to their children. But instead they left home when they grew up to marry outside of their home town, and now none of them make paper. In all of Kurodani there are only five or six people under the age of thirty making paper. Many expect that in ten or fifteen years at the most Kurodani may become another Yashiro.

### *Yashiro*

THE town of Yashiro lies barely one and a half kilometers along the river from Kurodani, tucked away deeper into the dark Tamba hills. It is known as a village from which all the young people have left—a town becoming a ghost town. The yellow wattle and daub houses are crumbling at the rafters, the gardens are overgrown, and the road leading up to the town is almost impassable; everywhere there is a sense of neglect and decay.

Yashiro was once a thriving papermaking town like Kurodani. Now papermaking is strictly a winter activity. The equipment there is old, outdated, and inefficient, but no one can afford to replace it. Of the twenty-eight families who remain, only five or six people make paper from start to finish, and all of them are over fifty years old. The young have never had, nor likely will have, any intention of learning this unprofitable craft. Almost without exception, those who have remained commute to Kyoto or Maizuru for work in cars bought out of their city job salaries. Knowing the hard work and sacrifice that a life of paper entails, they want no part of it.

All of Kurodani life is permeated with some portent of the demise of handmade paper. Recently, neighboring Hyōgo Prefecture requested an order of papermaking equipment, beaters and such, from the carpenters in Kurodani. Mr. Nakamura explained that just prior to this order some people came to Kurodani from this prefecture to receive three months of instruction in papermaking techniques. But it was not for the revival of washi in Hyōgo—it was for a demonstration of how paper was once made to be held at a new museum in what was once a famous papermaking town. I am reminded of a sad poem about the town of Sugihara in Hyōgo:

> The name of the birthplace of Sugihara paper
> still lives;
> But the papermaking houses—
> Now all have vanished.[4]

Since the Meiji period and the introduction of machine-made paper, Kurodani has always had to fight just to survive. Until about 1963 the papermakers here worked very hard. They rose at five or six and worked until midnight every night, barely making enough for food. The children were forced to quit school after their compulsory education, at fourteen or fifteen; the educational level here is not high. During the war, food production replaced paper production as the main line of work, partly because all the men and women were off fighting the war. Paper prices and sales fell two-thirds or more, and the types available were naturally limited. After the defeat, food was very scarce everywhere, so the villagers planted barley, potatoes, and the like instead of *kōzo*. But paper was in demand, even though its production was restricted by the government. Kurodani papermakers secretly sold their products at night to umbrella makers.

Why has Kurodani survived when so many other papermaking towns have failed? First, the community has kept a sensitive finger on the pulse of the market. When silk weaving was the major industry of this area, 40 percent of Kurodani's paper was related in various forms to the silkworm industry. Later, as this gradually declined, the town supplied paper for railroad baggage and made sandpaper. After the war, Kurodani papermakers experimented with recycling waste paper, since all paper was very scarce.

The village once produced *shibu-gami* (see p. 203) for umbrellas. Today it is still one of the papers most produced here, but it is mainly used in the dyeing industry and called *fuda-gami*. Almost all Kurodani paper is in some way related to art purposes.

Secondly, Kurodani has not sacrificed quality. Other towns tried to beat the competition by lowering their prices and short-cutting their work, but usually the results were disastrous. Not only was their paper weak and generally inferior, but ultimately they could not compete with machine paper. Kurodani realized long ago that it was foolish to try to compete. The people continued making strong, high-quality papers that were still in demand for certain specialized purposes.

Today Kurodani finds itself in an awkward state of transition. Whereas its paper is beginning to sell well, the very way of life that has made this success possible is falling apart. Modern life presents too many contradictions in the papermakers' life-style. On the one hand, the villagers cling to the old way of life and cherish it all the more in the face of encroaching modernization. They love the papermaking life. On the other hand, they find it almost unendurably hard; they want the freedom and luxuries that a good-paying job would bring. Some feel that papermaking holds the family together; and yet now it appears that it threatens to do just the opposite, as the young run to the cities for better-paying jobs and turn their backs on this centuries-old craft. And Kurodani is becoming increasingly perplexed by the small but steady attention the village is getting from the outside. Through the writings of Bunshō Jugaku and others interested in the future of washi, Kurodani and its paper are becoming more and more well known. It is sadly ironic that while the community must learn how to gently resist the threatening influence of tourism, it must also fight what is probably a losing battle in the struggle for Kurodani washi to survive.

*Ichibei Iwano*

SOMETIME in the sixth century, somewhere upstream of the Okamoto River in Echizen Province, Japanese legend has it, a goddess suddenly appeared from the skies. "The soil here is too poor for farming," she told the villagers. "But it is wonderfully blessed with clear streams." And so, casting her gauze veil upon the water, she imparted to them the knowledge of papermaking.

In the same town of Ōtaki one aging craftsman stubbornly clung to the most painstaking and conscientious ancient methods of making paper until his death in November of 1976. Ichibei Iwano's *hōsho* paper was considered the finest produced anywhere. Some of his paper covers the *fusuma* in the Katsura Imperial Villa. It is highly valued in Japan and abroad by woodblock artists for its robust and straightforward beauty, capable of enduring the numerous rubbings of a complex woodcut. This paper is so strong that, grasping it with the hands across the grain, a man trying with all his strength to pull it apart cannot tear it. If readers think this is an exaggerated claim, by all means test it. It is so durable that experts estimate his paper will last more than fifteen hundred years. Upon hearing this figure, Iwano quipped, "Well I've got some gall—when man himself can barely hope to live a hundred!"

At seventy, Ichibei Iwano had a startling intensity, energy, and desire to learn and improve his craft still further; the Japanese might say he never lost his *shoshin*, "beginner's mind." Always, he said, he tried to make paper putting into it all his energy and sincerity, without cheating or shortcuts. "It's not money I'm interested in," he said. "Although written differently, the word *kami* in Japanese means god as well as paper. Therefore I must approach my work with honesty—with my true mind and heart." Grunting and talking excitedly about his papermaking, he dashed from one part of his work area to the next in his great wooden clogs, leaving his visitors to scamper quickly along after him. His speech was friendly, open, brusque and full of colloquialisms. He regarded you directly, staring widely into your eyes. Even his slight braggadocio was charming and innocent, for he put the emphasis on the paper, not himself. My companion later described him as "full of prunes." Iwano described himself as "full of paper."

Iwano looked upon the triad of Japanese papermaking plants as a wise father looks at his children. To him, each has its own life and its own inborn proclivities—and it was his lifework to cultivate and develop the innate qualities of *kōzo* bark according to its true nature. "If I ignore those instincts," he said, "the fibers will respond negatively. In *kōzo*, if you seek perfection too much, the masculinity of the paper is weakened."

Iwano talked a bit about his early start in papermaking. He came from eight generations of papermakers. He began at the age of fourteen, learning from his father. Before he became famous, he used to go from artist to artist in Tokyo peddling his paper. Like the bleak, snowy country from which he came, life was austere. It was for this reason that he was not interested in training disciples, except his two sons and their wives. A life of papermaking is too hard for modern youth, he felt, getting up in the morning while it is still dark and working in the cold, cold water until late. He himself rose at 5:30 every morning, and went to bed at 9:00 P.M. He was sick perhaps one day of each year. His only recreation was sipping hot saké, and even that only at night—never at lunch—otherwise he could not work. He kept his workshop small, always working to perfect just one kind of paper, his *hōsho*. His output, therefore, was naturally limited; he never made much money. He was just proud to be a simple papermaker.

The Iwano family began concentrating on the exclusive production of *hōsho* during the early stages of the Meiji Restoration, after a fellow villager went to Tokyo and became a successful agent for woodblock papers. As Western pulp and machine papers began to overshadow that produced by hand, the paper craftsmen of Ōtaki (or Okamoto) decided to persevere in the traditional hand methods with more determination. Iwano worked in the spirit of this tradition; he stood out from his fellow craftsmen in his ultimate devotion to the making of quality paper.

When he was twenty, Iwano began pondering why even the finest paper made at that time rapidly turned into food for worms, while many very old papers bore the centuries so well. He began studying the oldest papermaking methods on record. Although he did not know much about science, he persevered in his study of paper chemistry. From his research he concluded that what the worms were eating was starch—produced in the paper when heat, during boiling, acted upon residual chemical matter in the fiber, turning cellulose into starch sugar.

These studies necessarily influenced his methods. For one, he kept his spanking clean workrooms at very cool temperatures, despite the personal discomfort, to preserve the freshness of the fiber. He insisted upon using only *kōzo*, which has the tough, coarse, and long fibers most suitable for *hōsho*. Many of his techniques maintained a closeness to nature and his materials. For instance, he used to lay the *kōzo* on flat stones in a shallow part of the river and tread upon it by foot, the old and primitive method for loosening the black bark. He used the very minimum amount of chemicals, which meant for boiling only a gentle alkali made of buckwheat, wood, or grass ash, or mild soda ash.

Iwano never bleached his paper; he obtained its magnificent creamy-white color by washing the fiber himself every day with his full arm submerged in icy, fresh mountain water. Although this washing decreases the bark yield, it is the process that dramatically reduces the starch in the fiber, he believed, and therefore adds hundreds of years to the life of the paper. Another surprising effect—his paper has almost no smell.

Iwano beat the fiber with long hardwood bats without the aid of machines—his is still perhaps the only sruviving workshop in Japan to do all the fiber beating by hand; the work is now carried on by his son, as it has been for the past several years. Hand-beating draws out the fibers, whereas machine-beating breaks them lengthwise so that they become "like the hairs of a cat," as Iwano chucklingly put it. He favored a propeller-type harrow for stirring the fibers in the vat, which was perhaps his only deference to mechanization. This was adopted into his studio only after experimentation with many new and traditional devices. The press his family continues to use is the most primitive kind imaginable—a great lever devised of a log set into the wall, which is gradually weighted down by hanging stones upon it. This also came after experiments with mechanical jack presses, which Iwano tried and rejected.

For drying, Iwano used gingko boards instead of the traditional pine and laurel woods, for the gingko can be sanded down to a very smooth surface, necessary for *hōsho* paper. Because of the scarcity of sun in his district through the winter, Iwano carried out his drying in a warm-air heated workroom on these boards—he refused to employ metal dryers. Every aspect of his work was in fact a result of experimentation, rejection, and finally adoption over many years. He never rejected a method or tool on the grounds that it was new or mechanized; rather he tried everything, and almost without exception found the old traditional methods the best. He sought good control over his materials and methods—but he instinctively knew that he must leave nature her margin of freedom in which to work her magic.

Iwano regarded his paper as a kind of offspring. After each sheet was finished and left his hands, he was very much interested in the direction its future life took. One of his greatest satisfactions was seeing and owning a fine work of art made with his paper. Often he made paper with the artist in mind. When he created *hōsho* paper for one particular artist who makes little animals from torn paper, Iwano purposely made the fibers longer, so that the paper was more furlike. When he created for woodblock artists, he kept in mind the color tone and degree of crispness favored by the individual. He was understandably proud of the famous artists who sought him out for his *hōsho*.

In March, 1968, Iwano was designated a "Living National Treasure." After becoming famous, Iwano found himself constantly entertaining visitors—which unfortunately took precious time away from his papermaking. Sometimes he unexpectedly received extravagant gifts from woodblock artists or gallery owners with no explanation—and then before long received requests from them to interrupt his heavy schedule in order to make them a special order of paper. These things saddened him. But his fame had its rewards too—it made it possible for him to obtain and afford the finest raw materials for his paper—a matter of great importance to him.

Iwano felt—"in his bones," he said—that the best season for his work was April, when the snow had melted from all but the most distant mountains. That was when he made the best paper, he did not know why—God's providence perhaps, he said—but he felt positive that was the very best time for him. This is about the same time of year as the cherry blossom season—but the work peak lasted longer, about forty days. It was the time when he felt in top physical condition.

Reflecting on the craftsman's life, Iwano once recalled "the old days" when there were no established vacations; people just worked until they were exhausted.

"Today life is so regulated by Sundays and national holidays—I kind of regret it. A real craftsman shouldn't work like this. Life is too businesslike it seems. Papermakers have always had to work under inhuman conditions. We are under constant pressure to make deadlines for orders—and yet so much depends on the weather! I find this sometimes depressing, and other papermakers do too. I'm concerned about the increasing lack of good materials, of labor, and of enthusiasm for the craft. On the other hand, there is a large demand for handmade paper today. That makes me happy, and I am encouraged."

*Eishirō Abe*

> I WOULD say that Abe's spirit, carved into his face over many years, is quite unique. I do not know to what extent his spirit may have been cultivated by religion or the books of saints; rather I think it must be the result of his being so absorbed in his work.[5]

In these words Takeo Kanaseki described Eishirō Abe, papermaker of the Izumo district in Shimane Prefecture, whose *Izumo mingei-shi* is known throughout the world. Many famous men have paid tribute to this remarkable man, who has devoted a lifetime of single-minded effort to the development of his paper. Not only is he constantly bringing forth new varieties of paper—he has also brought to life the nearly lost traditions that at one time produced eighth century *Izumo washi* in all its excellence. In recognition of his accomplishments, and to help provide for successors, Abe was named in 1968 a "Living National Treasure."

Some eight miles out of Matsue, near where the Izumo Grand Shrine mellows in gray dignity, lies the small village of Yakumo. Following a dusty road along the Iu River—a pilgrimage made by many lovers of washi, some renowned, most anonymous—one comes to a grand old thatched-roof house, Abe's home and workshop complex. One cannot help but be struck by the great beauty of this place, for here nature erupts in color, but as if from a core of calm. Abe himself has been compared to the beauty of Izumo, which has strength and constancy, and to the Izumo people, whose character blends a kind of bullish persistence with a deep appreciation for Tea and Tea taste.

Abe was born in 1902, toward the end of the Meiji period, when handmade paper was threatened with replacement by machine-manufactured paper. His father was a papermaker, as were his ancestors before him going back at least three hundred years. Although Abe eagerly wanted to continue his education, he was forced to quit school at the age of ten to take care of his sister and help with the family papermaking. But he studied at night school, and when he later took the military draft examination he was recognized as having attained a level equal to the third year of high school.

At the age of fifteen Abe entered a papermaking training school to learn newly developed techniques of hand-making washi. When he was twenty-one he married, becoming the adopted son of his wife's family, who were also papermakers. But the marriage was unhappy and after a year ended in divorce. This, Abe writes, affected him greatly at a particularly sensitive and impressionable point in his life. It was then that he became very determined to establish himself and develop a life through papermaking.

Abe married again in 1925 at the age of twenty-three. He and his wife worked hard making paper all day and turning the finished sheets into envelopes and cards at night, finishing work around eleven or midnight. Despite their hard work, they made less than three dollars a day. Times were depressed, and handmade washi was struggling to compete with machine paper. Around this time Abe was lucky enough to find a job teaching in a papermaking training school, which helped supplement the family income. In 1931 his life took a dramatic turn after meeting Sōetsu Yanagi while Yanagi was traveling in the area. Reportedly, Yanagi was electrified by the beauty of Abe's *gampi* paper, exclaiming, "It is the fulfillment of happiness in life."

It was largely due to the faith and encouragement of this man Yanagi that Abe became inspired to revive the once great *Izumo washi* to which he was heir by spirit and by birth. In 1955 he began a three-year study of some ancient Izumo paper from the Shōsōin Imperial Repository in Nara, paper that had survived over twelve hundred years. His purpose was to discover how it was made and of what materials, that he might produce a similar paper of the same superior quality. The secrets he discovered form part of his special papermaking techniques today.

Under Abe's guidance and his son's management, the Yakumo workshop produces much of Japan's finest *mingei-shi*, "folk art paper." Abe has developed about 160 kinds of paper, which is an unusually wide repertoire, and he keeps adding to this selection every year. He utilizes all three main papermaking fibers—*gampi*, *kōzo*, and *mitsumata*—sometimes mixing them in various combinations. But *gampi* paper is his specialty and the paper that made him famous.

Woodblock artist Shikō Munakata, who used Abe's paper, once wrote that there was a time when he used to get angry at Abe's first papers. He thought them showy and vulgar, containing too many things like tobacco and coarse fibers. But after meeting Abe and seeing his papermaking, Munakata's misgivings were transformed into a feeling of warm respect, for Abe's life was so wholesome and his attitude toward his work so serious. Eventually Abe developed a highly refined sense of beauty in his paper. He wrote:

> The heart of the man is very important in the world of crafts. You cannot make good things if you are out for money or trying to keep up with the fashions. It is necessary to have humility. There is a trinity of three things that go into the making of paper that will last forever—materials, water, and the man who makes it.[6]

Abe has placed much faith in nature and the natural. It is necessary, he says, to keep the fiber alive. His methods utilize as much as possible pure natural elements, such as the clear, cold water of the river nearby and the bright Izumo sun for drying the sheets. He avoids the use of strong alkalis in boiling the fiber and prefers preserving the natural color of the fiber to bleaching it. Often he dyes his papers, and nearly all of the dyes are traditional natural dyes—from indigo, walnut skins, earth oxides, oil of paulownia bark, and lichen to name a few, from the neighboring Yakumo fields and forests.

In his workshop the tools of papermaking are also of natural materials: wood, reed, bamboo, stone—even a camellia leaf is a useful tool, to smooth down the corners of thick paper on the drying boards. Nature vibrates here in Izumo, and Abe's use of it is clearly a studied one. Yet any observer can appreciate this man's insistence on using what is natural to produce papers so natural, wholesome, and clean in feeling. His papers are part of the beauty he has built around himself.

Abe's excellent taste lends grace to his entire living and working complex; this is a grace of clarity, harmony, and fullness of feeling. He has embraced folk art. In his house reside the works of anonymous craftsmen, alongside Munakata's prints, Serizawa's dyed textiles, and pots by Kanjirō Kawai, Shōji Hamada, and Bernard Leach—artists who are also his close friends. The second floor of his home houses a lovely one-room paper museum, crammed with interesting old paper goodies and relics. Framed on the wall is an often-quoted letter from Bernard Leach in which he speaks of the "friendly character and honesty of materials" of Abe's paper. "The hidden love and care involved silently request an artist or writer to employ the same human qualities when using it."

Abe has written and edited four books on his papermaking: *Izumo no kami* (*Hand-made Paper of Izumo*), *Kamisuki gojūnen* ("Fifty Years of Papermaking"), *Washi zanmai* ("Devotion to Washi"), and *Izumo washi-kan* ("The Izumo Washi Collection"). In *Washi zanmai*, he wrote of himself:

> My life has been led by tradition. With great predecessors such as Yanagi, Kanjirō Kawai, and Shōji Hamada and with many friends and acquaintances, it seems as though I have been a young tree growing in a natural environment without ever being aware of my own growth.[7]

And as a friend of his put it, the garden tree has grown into mountain timber.

*Seikichirō Gotō*

SOMEONE recently described craft as doing something over and over again, but art as one's expressing oneself for all time. By this definition Seikichirō Gotō is an artist, but one whose artistic flights always return to folkcraft traditions. He is the only Japanese papermaker who is also a noted artist-craftsman. For him, creating is as necessary as breathing. New mediums excite him, and he is versatile in many. He is a stencil artist and printmaker, a dyer, a weaver, a *sumi-e* artist, the developer of *inden-shi* (lacquer stenciled paper) and *kinkara-washi* (relief lacquered paper), to name a few. Years ago he invented a kind of bamboo paper called *chiku-shi*, which produces a beautiful tone in high fidelity speakers. A famous Japanese company has been using it in their commercial sound systems for years. Although Gotō's involvement in paper and papermaking came late in life, he is an active and accomplished paper craftsman.

Gotō was born in Ōita on the island of Kyushu in 1898. He graduated from Kyoto's Kansai Fine Arts Institute. Then, in a highly unorthodox move, traveled to India to study textiles and, in particular, printed fabrics. He studied in India for four years. Upon his return he moved to Chiba Prefecture, not far from Tokyo, where he met Sōetsu Yanagi. Gotō soon began studying a kind of textile dyeing that was being revived and further developed by Keisuke Serizawa under Yanagi's direction and encouragement. This dyeing process, called *bingata*, is a traditional Okinawan craft using stencils and paste resist to print beautiful and often complex patterns. World War II arrived—the production and distribution of cloth was limited and restricted, so the group experimented successfully with dyeing handmade paper. A new craft was born.

Gotō had long had an interest in papermaking, and now that interest grew even more profoundly. He was quite taken by Bunshō Jugaku's *Kamisuki-mura tabinikki*

("Diaries of Pilgrimages to Japanese Papermaking Villages"). He decided to make his own pilgrimage, and with great hardship traveled all over Shizuoka Prefecture visiting paper villages while the war raged on around him. This eventually resulted in a book of his own describing his travels. After the war he moved to Fujinomiya, in Shizuoka Prefecture, his wife's hometown. There he decided, at Yanagi's urging, to devote himself to papermaking. The beginning years were extremely difficult. All of Japan was close to starvation, and the Gotōs were no exception. Yanagi used to visit them bringing gifts of food and words of encouragement, kind acts that the beginning paper craftsman never forgot or failed to appreciate in the years to come. Paper scholar Bunshō Jugaku was also a great encouraging influence on Gotō. Dr. Jugaku wrote the introductions for some of Gotō's books, books he continued to produce on his own paper and containing his original stencil prints.

Gotō does not really consider himself a folk artist—his technique is too sophisticated and his tastes too cultivated and trained. But his heart goes out to folk art and the beauty of natural materials as they are transformed in the hands of natural and simple people. Neither is money or fame important to him; he has pursued neither, though they were within his reach.

One of Gotō's greatest heroes is Yanagi, because, Gotō says, "he really loved and found beauty." Gotō believes that the real folk art movement is not a society at all, but people's ability to find beauty in the everyday things of life. "Use is of prime importance," says Gotō. "A rich man will collect folk art and put it on a shelf, and indeed it looks beautiful, but when it is not used it lacks all strength. Folk art should live and function in daily life." If a husband does not work, says Gotō, his wife will have bad feelings toward him, and vice versa. "It's the same with crafts—they become more beautiful with use, just like a good wife. When something is wholesome, when it fills its role, it is beautiful. That is the beauty of folk art."

Gotō feels that paper made by hand has some mysterious beauty that machine paper lacks. "Hands have the possibility to make what the machine cannot." His own paper has much character. In the main he uses *kōzo*, but sometimes *mitsumata* or a combination of the two. Generally he does not discard the green cuticle of the *kōzo* bark, which gives his paper a somewhat gray-greenish cast. His paper is quite strong, with fibers noticeably swimming on the surface. Strength rather than refined texture is his main concern, because the paper is mainly for print-dyeing and *kinkara-washi*.

To a great extent Gotō's prestige has been built by his revival and further development of an old art, which he renamed *kinkara-washi*. In the late 1800s, a gorgeous, thick paper with birds and arabesques in relief appeared in Japan. This leatherlike substance was made by impressing thick, damp handmade paper into fine patterns carved out of large cylinders of cherry wood and coloring the paper with lacquer. Its name was *kinkarakawa-shi*—literally, "gold China leather paper." Its origins may go back to the Edo period, when Buddhism discouraged the killing of animals, especially draft animals such as cows, oxen, and horses. A type of thick and usually embossed paper came to be made. Supple and strong, it resembled leather. This paper was formed into boxes, purses, briefcases, and tobacco pouches as well as many other items, then lacquered.

*Kinkarakawa-shi* made its Western debut at a world exposition in the last century at the suggestion of, legend has it, an Englishman nicknamed "Old Watch" who was a railroad conductor in Japan at the time. Soon after the exposition, it was exported in quantity to Europe and America. Reputedly some of it was used as wallpaper at Buckingham Palace. *Kinkarakawa-shi* enjoyed great popularity for many years abroad; but eventually it began to be made with poor materials, soon lost its shape, and after the beginning of this century production stopped.

Gotō knew about and admired the old *kinkarakawa-shi*, and for years sought in vain the beautiful, intricately carved wood rolls. He even tried to carve the wood himself, but found it too much work. One day in Tokyo in the 1950s he ran into a second-hand goods shop to get out of the rain—and there found fifteen of these great rolls. He bought all of them on the spot and thereafter worked to redevelop the art. He calls his product *kinkara-washi*, taking the old *kinkara* and adding the word *washi*. He makes the sturdy paper used for it himself and coats it with persimmon tannin and lacquer after it has been pressed and dried. The making of *kinkara-washi* gives Gotō a deep satisfaction, for, in his own words, it "preserves washi forever."

*Kinkara-washi* is used as fronts for cabinet doors, boxes of various sizes, simple decoration for pottery, trays, and, framed, as hangings. Gotō is the only one in Japan today practicing this craft.

Over the years Gotō has published eleven books, including *Nihon no kami* (*Japanese Paper and Papermaking*), 1958–60, in two volumes; *Kami no tabi* ("Paper Travels"), 1964; *Shifucho* ("Book of Paper Development"), 1954; *Washi-inden* ("Lacquer Stenciled Paper"), 1957; *Kinkara-washi* ("Kinkara Paper"), 1971; and *Kami o suku ie* ("Papermaking Houses"), 1972. Many of the books are printed on paper of his own manufacture with original hand-stenciled prints of his own design—tremendously delightful books that are now collectors' items.

In the mid sixties Gotō traveled to China with a group of Japanese papermakers by invitation of the Chinese government to observe hand papermaking and enter into dialogues on the revival of the craft there. In 1971 he was awarded the Fifth Order of Merit by the Japanese government for his artistic contributions to the nation.

NOTES

half-title: Tokutarō Hanada and Kiyofusa Narita, *Kami-suki uta*, 2 vols. (Tokyo: Seishi Kinenkan, 1951). (group trans.)

1. *Tesuki washi taikan* (*A Collection of One Thousand Handmade Japanese Papers*) (Tokyo: Mainichi Newspapers, 1973–74). Some English text.

2. Sōetsu Yanagi, *The Unknown Craftsman*, trans. Bernard Leach (Tokyo: Kodansha International, 1972).

3. Hanada and Narita, *op. cit.* (trans. by Yasuo Sasaki)

4. *Ibid.* (trans. by Yasuo Sasaki)

5. Takeo Kanaseki, "The Mask of Abe," in *Izumo no kami*, ed. Toshio Ikeda (Iwasaka, Shimane Pref.: Eishirō Abe private pub., 1952).

6. Newspaper interview, 1973. (trans. by Louise Picon Shimizu)

7. *Ibid.* (trans. by Louise Picon Shimizu)

*It is said that paper is made by people, but it would be better to say that the blessedness of nature produces paper.... Those who study washi study at the same time the depth of nature.*

Sōetsu Yanagi

# The Simplest of Substances

Certain Japanese, whose lives have long been intimate with paper, speak of handmade paper as a living substance. What is washi? To what extent is paper "alive"? What is so special about this seemingly ordinary product that stirs men to write essays and entire books on its character and beauty? What impels so many to breath quiet reverence when they hold a single sheet of well-made paper to the light?

When serious, thinking men love the same thing and talk reverently of it, one is impelled to listen to them. Undoubtedly I have tried to be open to whatever insights a knowledge of washi has offered, but I have not had to search too hard or too deeply to find them. Not only my teacher, Seikichirō Gotō, but also Japanese craftsmen, artists, and scholars such as Keisuke Serizawa, Bunshō Jugaku, Sadao Watanabe, Shōji Hamada, Shiko Munakata, Eishirō Abe and the English potter Bernard Leach to mention a few—all of them have paid homage to washi. Sōetsu Yanagi, who founded the folk art movement in Japan, spoke of paper's "soul"; to papermaker Ichibei Iwano, paper was his "offspring." This chapter takes on what is probably a presumptuous task in exploring paper's inner nature as I see it—that is, paper as it acts as an interface between an inner sense of beauty and its outward expression in the world; between artistic creation and man's self or truth as he realizes it—a place where nature and the human spirit meet. The qualities of washi are in every sense an expression of Japanese culture and craftsmanship as well as an expression of this country's ideals of beauty and intense feeling for nature.

*Nature and Craft*

As always in Japan, everything of value starts and ends with nature. Nature is the great teacher. One cannot escape from her beauty any more than from her fickleness or the changes and moods she creates within one. The craft of papermaking starts and ends here.

The Japanese papermakers have created an intimate relationship with nature—they have succeeded in living inside her. Nature is not regarded as something to be enslaved and dominated—in fact she cannot be, for the power and drama of the Japanese climate forbids it. She is the generous provider as well as the devastator—what she gives she is just as likely to snatch away. The papermaker's work has always been marked by a sense of urgency, in meeting deadlines and in working around the fickleness of weather and the quick spoilage of materials. Nature chains the papermaker at every turn, for his gainful labors are dependent upon manifestations of her goodwill—her sunny days, her harvest of wild and domesticated bast fiber plants, her winds that dry paper crisply. But the chains are light because the paper craftsman loves and respects nature as much as he loves and respects his craft.

The Japanese are resigned to nature's irrationalities, but have also learned to know her well, to anticipate her facile mutabilities and to flow with them. The papermaker puts his ear to nature's heartbeat and his finger to her erratic pulse. He accepts her heavy conditions, but works hard to keep pace, ahead of them. Just to try to make good paper is his purpose; that alone is quite enough. Papermaking is labor, but it is also a refined skill requiring years of practice, patience, and sensitivities to materials and their capabilities; such sensibilities develop from translating nature into paper craft. The worker knows that it is only nature that can give his paper vitality. The astute papermaker does not want to carry on his craft outside nature, above and beyond her realm, as with machines; that would be at once irrational and artificial to him. He wants to absorb nature into his work. Every sheet of paper must bear the look of her touch, and that is evidence enough of his craftsmanship. The craftsman, lettered or unlettered, creates beauty because, as Yanagi said, "he rests in the protecting hand of nature."[1]

Folkcraft is very much an outgrowth of rural, simple life and its poverty, a life in which people were required to make their own furniture, utensils, and clothing—in fact to make most everything essential for living. And so craft has substance and a special validity because of its practical nature. Because it is a form of physical labor, of work, and because it may also be a livelihood, it is ever an expression of the craftsman's sincerity, his unique and serious relation to life. Craft has a spiritual element because its roots are buried in the earth, and its fruits the tools of day-to-day life. Craft creates a bond between nature and man, between the laws of earthly existence and man's idea of beauty.

The Japanese papermaker still draws his sustenance from the earth. Into the soil he sows and from the soil he harvests his raw materials. He steams off the bark in a steamer built in the open fields. He soaks the bark in the rivers and streams weighted down by stones. His tools are handmade; he instinctively reaches for the natural in fashioning them—a split bamboo for a pipe channeling water from a nearby stream to his house;

brushes made of straw, palm fiber, or horsehair for smoothing the paper onto the long drying boards; a glossy camellia leaf to firm down the edges of a drying sheet; bamboo or reeds for making the paper screen; stones and logs for a press to remove water from the sheets. In his work, a piece of bark fiber substitutes for a suddenly needed cord; a twisted piece of his own paper serves as a string for binding the finished sheets. These bits of nature, the papermakers feel, suit the character of washi, are its proper companions.

In Japanese country life, craft and natural integrity are not separate and independent aspects of existence; on the contrary, they are all related threads that form patterns of living. When a man's work is not in harmony with his life or with nature, it is evident in what he produces. This is especially true with papermaking. Paper reflects the mind of the maker, and both "Living National Treasure" papermakers sensed this. Ichibei Iwano told me that he could look at paper and see at once if the maker was quarreling with his wife. Eishirō Abe wrote, "When I take a sheet of paper in my hand, it tells me about the earnestness of the family that made it. If someone in the family did not keep pace with the others or forgot a task, it shows up in the paper, which has a feeling of something missing."[2]

Seikichirō Gotō once said to me that he can look at paper and see if it is good. "Seeing good paper is like seeing good people," he said. "Bad people don't look you straight in the face. But good paper—it will meet you squarely in the eye. Like food, you just have to try it out in order to judge its qualities. Yet, if paper is made honestly, even by a bad person, something special happens—the paper will come out well." It is as if in its sincere making, the papermaker carries on a communion with nature that is beyond questions of good or bad.

Every material found in nature—clay, wood, fiber, stone, straw—all have their characteristic proclivities. The artisan learns to become sensitive to these proclivities and potentialities in his chosen material. The raw material of paper has its own instincts. To go against the innate "will" of the paper fiber is of course possible, and the results may be quite novel—but ultimately it will not "work." The fibers' instincts are too strong; nature, as it were, takes offense.

How often it seems that man has forgotten the unwritten law, especially true for the arts, to never stray too far from nature. It seems to me that any object of true beauty must first capture the essence of its own raw materials and display those inner qualities strongly, simply, clearly, and naturally. Giving the highest compliment to Abe's paper in a letter, Bernard Leach wrote that it was "so close to the fields and fibers in character." Is it nature that we see flowing out of washi, or is it humanity, or both?

*The Place of Tradition*

"Why is washi so wholesome?" asked Yanagi. "When we try to figure it out, we cannot help but think that it is because nature is paper's mother and tradition paper's father."[3]

In the East, where tradition commands respect if not reverence, Japan has set a miraculous sprint record for change, not only in adopting the modernization of the West but by becoming a pioneer in her own right. Changes in the arts, however, have moved somewhat less spectacularly. Like any other nation, the Japanese, too, get excited about artistic revolutions, and they have a very definite and intense love of Western art and music. What is happening today in Japan is a struggle for the survival of the traditional roots and the cultural attitudes of appreciation that they have fostered over the centuries. Renewed interest in and revival of dying arts and crafts by the young does appear to be occurring more frequently today, but it is too early yet to gauge how long and how well these interests will hold or what permutations will occur. Older Japanese know very well that traditions, when they do survive, survive for a reason. They are the tried and proven forms, the knowledge, methods, and objects that man has tested through generations of experience and found to be most valuable and enduring.

It is interesting, I think, that Ichibei Iwano, renowned as one of the greatest Japanese papermakers, discovered that the traditional methods and techniques passed down to him from many generations of family papermakers were almost without exception the best. Questioning everything that was passed on to him in a never-ending effort to improve the quality and permanence of his paper, Iwano studied, experimented, and studied more, to finally make but one rather minor mechanical change in his equipment. In every other aspect of his work he returned to the pure traditions passed down to him. But he needed to know for himself their true value.

Traditions become stagnant only when they fail to bear the necessary mark and changes of each generation through which they pass. They become withering shadows of what they originally stood for if their truths are not rediscovered by successive generations, who must breathe new life and meaning into them for their principles to remain significant. This means that tradition, to continue as tradition, cannot allow the threads of its structure to stiffen and petrify. Tradition is a strong, resilient warp through which the new and living vision of each generation is woven. What makes forms true and valid to begin with must be experienced again. Each papermaker must constantly go through the initiation of essential meaning in his labor. Every time he makes paper he is the first man to make paper, and every new sheet is his first sheet.

Yanagi once said that the greatest achievement by the arts is the discovery of law. In the crafts, tradition is a kind of law that does not restrain, but leads ultimately to absolute freedom. Freedom of expression as we in the West tend to understand it too often implies spontaneous expression of our personal idiosyncracies, of what we think makes us unique and different. But what kind of freedom is this really? More often than not it is but an expression of personal limitations, of confusion, of spiritual vacuity. For as long as a craftsman ignores the value of discipline and training, for as long as he is caught up in emphasizing his ingrown involvement with self, he cuts himself off from something greater than himself—from his relation to the universe, as it were—and his work remains adolescent.

Traditions in craft are not merely techniques. They contain hidden lessons. In craft the little self must be held a bit at bay. Tradition is the greater wisdom. It provides for us the nourishment and the structure through which our smaller selves begin to sense their creative roles within a grander realm of nature. The papermaker, through his digestion and absorption of all that the traditions of his craft offer him, lets the materials speak; he leaves "intention" behind. Freely he allows the materials to be what they are and to express themselves according to their own proclivities, not according to his own willfullness or desire to "express himself." Then, his body, his mind, his heart truly are in total harmony with materials, work, and product. Then the papermaker embraces "washiness"; he becomes the paper.

One secret of beauty is irregularity. As Francis Bacon said, "there is no excellent beauty that hath not some strangeness in the proportion." True beauty takes us to the irregular, but the difference between natural irregularity that arises from obeying the laws implicit in the materials and irregularity created willfully by the confused artisan is an important one. The craftsman cannot force freedom; when he heeds the laws of tradition as well as the lessons of experience and the instincts of the materials, yet allows nature to wield her special magic—then he achieves freedom in his work, and beauty and wholesomeness naturally follow. When he and the work are one, the work expresses itself.

In papermaking there is not much room for creativity, but there is more room for honest work. So honest work becomes the greatest responsibility of the conscientious papermaker. As the clear instincts of the fiber are sincerely followed in its manufacture, the individuality of the maker is submerged; then paper becomes true paper.

*Washi and the Japanese Idea of Beauty*

THROUGHOUT Japan's long history, since the inception of the craft of papermaking, handmade paper there has been reverentially treated as something pure, chaste, unaffected, of strong and wholesome character—in fact, a worthy offering to the divine. The beauty of paper is the same as that found in all folkcrafts. The qualities of commonness, naturalness, irregularity, simplicity, and warmth inspire affection for washi. As Yanagi said, "the humble are receptive to love."[4] If one becomes absorbed looking at washi, allowing its nature to express itself, one begins to see that it is not only its physical beauty that one understands but that one also can get to feel the whole of it, to know the labor behind its birth and the simple dignity of its evolvement, the "flowering" perhaps of its plant nature—to perceive its "soul." As the twentieth-century Japanese philosopher Tetsurō Watsuji said: "For although one only looks at the exterior, one feels that one can see all the interior."[5]

Around this core idea revolve the Japanese ideals of beauty, a beauty of "inner implications," in Yanagi's words. "It is not a beauty displayed before the viewer by its creator; creation here means, rather, making a piece that will lead the viewer to draw beauty out of it for himself."[6] It is a beauty of suggestion rather than of explication. "Not to display, but to suggest is the secret of infinity," Kakuzō Okakura said. "Perfection, like all maturity, fails to impress, because of its limitation of growth." Moreover, beauty, he said, "or the life of things, is always deeper as hidden within than as outwardly expressed . . . ."[7] It is this concept that impelled the Japanese to work the most intricate and exquisite gold lacquer into the inside of a box rather than onto the outside.

This kind of beauty, a beauty that makes an artist of the viewer, stems from Zen Buddhism, particularly Zen's outward manifestation of its aesthetic ideals in what has been called the Way of Tea. Before the dramatic changes affecting aesthetics and taste brought on by Zen and the Tea cult, Buddhist art generally had become increasingly sterile. Like the religious thought behind it, early medieval Buddhist art in Japan had grown shallow and gorgeous; it aimed at impressing the masses with its intricate colors and patterns rather than inspiring spiritual insight. On this Zen turned its back and, rather, sought to discover the profundity hidden within the real and the simple. Zen people attempted more direct interaction with and expression of the world around them. By going back to essentials, by really looking at nature and handmade things, the Zen monks discovered an inexpressible beauty in the irregularities and unaffected expressions of such objects as paper. I think it can be said that the most beautiful, the most ingenuous of these papers reside in the heart of Zen.

The core of both Buddhist philosophy and Tea alike is embodied in the word *yūgen*. By strict definition, *yūgen* means profundity, mystery, or abstruseness; by implication it means to feel or touch the mystery of things, to be close to their truth. It expresses a quality that "moves perpetually in its stillness." Thus beauty of this kind has a dynamic quality, beseeching the viewer to constantly distill ever new meaning from artistic expression, to develop an increasingly pure refinement of feeling. The sincere seeker encircles his life with beauty until finally beauty enters and rests within him.

The somewhat ritualized practice of drinking tea was brought to Japan in the thirteenth century by Ch'an (Zen) monks from China, where Ch'an meditants used the infusion as a stimulant. It was the Japanese who took the simple tea-tasting games and developed them over a period of time into an elaborate ritual vastly creative in its impact on a culture. In fact, culture itself blossomed from this cult of Tea. This was one activity that covered all activities, so central was it to the source of all being, and to the core of Zen. Simplicity and humility, sincerity and restraint, gracefulness and economy of movement, harmony and quiet—all mingled with a profound love for nature, which found mysterious reserves of beauty in the impoverished and the commonplace. Tea ceremony moved out from the monasteries and into the lives of the people, who, in the centuries that followed, found in its movements and laws a deep spiritual base, a retreat into nature where they could still the clamor and suffering of the outside world. As the cult of Tea grew to maturity, a universe of aesthetic sensibility opened up.

By the late sixteenth century, Tea had become a private tabernacle of all Japanese. Vanity, hostility, and pride were to be left outside the garden gate as one approached the place of Tea. Men and women alike, soldier, servant, and statesman were equal under the Tea room roof—only the humility, sincerity, and spirit of peace a person brought along were the measure of his worthiness. Unquestionably Tea too often became formula and sterile—a platform for showy displays of all manner of poses and silliness—the antithesis of its original aim. But at its best Tea was and still is an extraordinary ritual in what it conveys to man about the art of life. It was here in the Tea rooms and Tea huts that the great masters, usually Zen priests, posited those standards of taste, in life as in art, that survive today as being especially "Japanese."

The early masters chose their Tea utensils from objects made and used by country folk as part of their subsistence. Because folkcraft objects lack artfulness; because they are rustic in character and mellowed with age and use; because they have an unhesitating naturalness; and most of all because their rough beauty is an approachable beauty, familiar and humble, that inspires compassion and begs to be used—for these reasons they received the affection, the allegiance, and the greatest esteem from the men of Tea.

The Tea masters chose washi for use in the Tea ceremony undoubtedly because the laws that govern the beauty of Tea implements govern the beauty of good washi as well. Even newly made washi has a mellowed quality to it; its color is creamy white, natural. Fibers dance on its face like clouds in a broad expanse of sky. The small pieces of bark in *chiri-gami* remind one of the skin of an old man or woman, slightly dark in tone and flecked with age spots. Washi is egoless, passive, still. It has a humble beauty. Good washi stands on its own, totally without artifice, expressing nature much more than the personality of the man who made it. Its imperfections attest to a strong and vital character, such as the unshorn deckle edges and random pieces of coarse fiber floating on the surface. It is simplicity itself.

### *A Language of Beauty*

THE cult of Tea created a new language of taste, a vocabulary of beauty and feeling that invested old words with new subtleties of meaning to suit the needs of this now highly refined aesthetic. Words such as *shibui*, *sabi*, *wabi*, *aware*, *musō*, *yūgen*, *kareta* and more—all have deep nuances of meaning that are extremely difficult to fathom, much less define. The very existence of such words implies the existence in turn of a whole world of sense and feeling hardly imagined in the West, a universe of significance lying in the shadow of the most apparent phenomena.

*Shibui* is a word used to describe a quality of mood, or taste. It has three dictionary definitions that, although they seem unrelated, are various nuances of one complex meaning: astringent; glum or sullen; and quiet, sober, tasteful. It implies restraint and calm. *Shibui* beauty is the beauty of Tea, the beauty of "inner implications." Not a static beauty, it changes with the viewer and according to what is placed in contrast to it. *Shibui* beauty has a profoundly stable and quiet feeling, as if it was rooted in the earth. It incorporates, says Yanagi, "ideas of simplicity, quietude, propriety, spontaneity, and the like, and holds the beauty of nature and health in high regard."[8]

Originally meaning "desolate" and later "aged," *sabi* eventually came to mean that which is lonely, bare, and cheerless, but also ripe, deep, approachable, mellowed with age and use, faded, traditional, quiet and settled. *Sabi* also implies the pleasure in discovering anything imbued with these qualities. Nothing showy, gaudy, or overwhelmingly dazzling can have the quality of *sabi*.

A common simile used to express the feeling of *sabi* is seeing an old warrior, skilled, dignified, and decorated, fighting nobly and well in battle; his age stands in sharp contrast to his deeds. *Ibushi-giri*, tarnished silver, is considered the "hue" of *sabi*. To express *sabi*, Zeami described snow falling into a bowl of oxidized silver. Zen philosopher Shin'ichi Hisamatsu suggests that it is "the quality of being seasoned enough to appear tranquil, serene, antique, and graceful."[9] Stones may have the quality of *sabi*. The depth at which one places them into the ground in a rock garden enhances or diminishes the garden's spirit of *sabi*. *Sabi* locates the center of stillness in the world. Like the Tea room it typifies, it is a retreat into serenity.

The more profound meaning of the word *wabi*, as known today, grew to maturity in the latter part of the sixteenth century when Sen no Rikyū established the *wabi* style of Tea ceremony. Originally meaning lonely, forlorn, and miserable, it came to describe some object that gave the feeling of "fulfilled poverty" or to describe the feeling of fulfilled poverty itself. *Wabi* is that which is poor to the point of being in want. Although less refined than *sabi*, and lonelier, more rustic, it retains the *sabi* qualities of simplicity, naturalness, and tranquillity. In fact there are "two minds" to *wabi*: on the one hand, refinement, fulfillment, and balance are desired; but on the other hand, imperfection, lack, and imbalance are considered more virtuous, and so—in compassion—accepted.

*Wabi*, then, is not only a standard for judging beauty in nature and in things, but it is also seen as one way to approach life. Here there is a contentedness in bareness and in doing without, such as in the limitations imposed by traveling or leading a wanderer's life. Beauty is found in extreme simplicity and the simplest type of existence, which implies living close to nature. If you are not starving, you have enough to eat. If your house keeps out the rain, that is enough. In this sense *wabi* means not only to live in such a manner, but to gain peace of mind and contentment realizing the beauty of such a life. For although one may be impoverished, one does not think of it as impoverishment.

"All things are beloved for their imperfections," said Ruskin. *Wabi* beauty always contains something that is unbeautiful. Its beauty is rough and underdeveloped but calm and harmonious. Because it lacks shine, it possesses depth. Its irregularity has balance, however, for "what appears to be symmetry to the ordinary eye is of a lower kind, but that symmetry which may be found in apparent irregularity is of a higher kind."[10] *Wabi* beauty would draw us into it, as it would also pervade *us*. Although it may be new, an object imbued with *wabi* appears to contain an old soul. *Wabi* beauty has a humility and sincerity we revere.

Tea taste becomes what it strives to be when it achieves the feeling of *wabi*, and to some extent *sabi*. In the humble form of Tea ceremony, everything is chosen and arranged to give a spontaneous suggestion of *wabi*. That is why so many old folkcraft artifacts and utensils are said to have *wabi*.

Some kinds of washi and items made from it are imbued with qualities of *wabi*. I think that the paper made today in the Mt. Kōya area overflows with it. This paper is made under intense hardship by two lone women, one quite old. Their paper is simple, crisp, and strong in character; the color is dark, and its surface is flecked with stray bark. It is so modest and plain that it could be called impoverished. I think it could well take its place in the *wabi* style of Tea ceremony.

Wearing clothes of handmade paper has always suggested a *wabi* quality of poverty in Japan. The poet Bashō was quite fond of *kamiko*, paper clothing. On the practical side, he must have appreciated the paper kimono as a good windbreaker and insulator against the cold, but perhaps it was the spiritual and emotional implications of *kamiko*—its *wabi*ness—that made him love it. This is the feeling implicit in his haiku about *kamiko*.

Bashō often slept in the open air, perhaps on piles of grass or straw. Awakening in a field early one cold autumn morning on his travels, he wrote this haiku:

> *Kamiko ni mo shimo ya okukato nazetemishi*
>
> Upon this very robe of paper—
> Oh! I touched some frost.[11]

What are these mysterious qualities of *wabi* and *sabi*? If you should ask a young Japanese who has thought about such things, chances are that he will refuse to define these terms, excusing himself by virtue of youth. "Ask someone over fifty or sixty," he will say. What he means, evidently, is that one cannot know *wabi* and *sabi* without experience. It would appear that, to understand them, we must give life time enough to properly deepen and mellow us.

*Wabi* and *sabi* have no particular color, no size, no shape; there is nothing about them that is static. On the contrary, they change from time to time and place to place, according to how each may experience them. Although they seem to represent certain qualities that are more or less commonly agreed upon, the actual experience of *wabi* and *sabi* is always highly individual and personal. It involves feeling keenly about life and everything around us.

By the very existence of such terms we are enticed to discover their expression in objects and events, to strain our intuitive faculties trying to understand the mysteries of which they hint—and thereby create an inner room for them. Knowing that they are something profound and hard to discover, we may find ourselves looking at everything in a new way, approaching all with our total self and guarding against discovering *wabi*, *sabi*, or *shibui* too easily, too quickly.

Sōetsu Yanagi, in particular, saw and wrote with great feeling about the qualities of washi. In "The Beauty of Washi," Yanagi talks of Japanese paper as of an intimate.

> There is no simpler material imaginable. But when one looks at it, one is captivated. Handmade washi is always filled with charm. I stare at it; I touch it with my hands—and I am filled with a satisfaction that is difficult to express. And the more beautiful it is the more difficult it is, I find, to make trivial use of it. Washi would be insulted with other than the best of calligraphy. It is a marvel just as it is. When you think about it, isn't it strange? It is only paper. However, it seems that its beauty is enhanced by its very plainness. Good paper inspires us to dream of good things. Therefore I think about the nature of paper and I think about its fate.
>
> As always, I meditate on the source of its beauty. Its beauty must be the beauty of quality. Its original quality being good, when it is worked on by hand, it is reborn into the best of paper. What do we mean by "quality" though? It is washi's blessing from heaven. The more this blessing is effused throughout it, the more beautiful it is. Saying this, the riddle is solved.
>
> Why is it that when paper is made by hand it becomes warm? Why is it that the natural color cannot go wrong? Why is it that when it is dried in the sun, the tinge of the paper is serene? Why is it that when it is dried on boards it comes out better? Why is it that the winter's water preserves the quality of the paper? Why is it that when the edges of washi are left uncut, it becomes more refined? The truth seems to be quite obvious. It is because when paper is made by hand its heavenly blessing is most warmly expressed. Nature then shows her depth without hiding anything. When the power of Nature is most strongly felt, any paper becomes beautiful. It seems justifiable to think of the beauty of washi in this way.
>
> There is no "I" in paper. As a result there is no one who hates it. In this lies the amiability of paper. Those who do not reflect on paper will probably be indifferent to it, but those who become close to it will most likely feel a bond to it from which it will be difficult to part. Never once have I shown paper that

I love to someone without that person feeling joy from it. Those who look at it once always take another, closer look. Good paper inspires love. Good paper will make us feel deeply a reverence for nature and a passion for beauty.

In washi there is also a kind of joy of discovering Japan. One could not find paper like this in any other country. Washi makes Japan more beautiful. When in Japan it would be a shame to forget washi.

How much paper is consumed by a country is supposed to show the degree of that country's civilization. But this refers to quantity of paper; I would rather it meant the quality. We should measure the degree of feeling—is there any connection between bad paper and good culture? Especially, what kind of paper is selected for stationery or what papers for books in our everyday life? Just from knowing the quality of the commonly used paper we can guess the norm of the nation. Those who ignore washi ignore beauty as well.

On our left side we can see incredibly poor paper; on our right, paper that is infinitely beautiful. Which is to be selected depends entirely upon the possessor. Possessions and the possessor are not separate. Man should be a good selector.

At present, people make light of paper. This is because of the increase in the amount of paper that can be used wastefully. Or, rather, it is more appropriate to say that the desire for quality paper has decreased. But how can we find happiness in a kind of life in which paper is used wastefully? I would like to avoid this kind of attitude toward washi as much as possible. It is not desirable either morally or aesthetically, for that kind of attitude lacks the spirit of gratitude.

What has caused this unhappy situation—the decline of washi? Western paper has taken over. Paper made in the pure Japanese fashion is never ugly; but most people consider it unsophisticated, and so "improvements" were too quickly adopted. As a result, however, the intention behind these "improvements" has greatly worsened the quality of washi. The reason why the washi made today is of such poor quality is that it has been made ignoring its historical traditions. Besides this, the spirit of profit has thrown away the beauty of washi, and without regret. Why haven't papermakers respected their heritage and devised new kinds of washi? There is no safer foundation upon which to stand than upon tradition. If tradition was respected, Japan would have no competitors in the field of paper.

I would be accused of exaggeration if I were to say that any kind of washi is beautiful. I will then say, rather, show me something ugly in washi. It is impossible. The process by which it is made promises just that much beauty. So paper made even today, when made in the traditional way, is without defect. No paper made in the traditional way is "unhealthy." In the traditional process no imitation is allowed. Handmade paper that bears the weight of history has no defect. The only question left is which washi is more beautiful.[12]

### *Calligraphy and Ink Painting*

At this point it is undoubtedly good to stop talking of washi as an isolated material and instead to look at its role as a receiving surface, as in the arts of ink painting (*sumi-e*) and calligraphy (*shodō*). The quality and nature of the paper used are as important as the ink, inkstone, and brushes. A sheet of white washi has qualities of stillness and emptiness of which men of the brush are acutely aware. They see paper not only as a blank slate for creation, but as a potential field upon which dynamic, internal forces will play. By limiting himself to black ink and white paper, the artist forces himself to deal single-mindedly with the problems—and infinite possibilities—of interacting positive and negative spaces.

Reaching a high point of development in China under the Ch'an Buddhist masters, ink painting traveled to Japan with returning Zen monks. Calligraphy already had a long history in Japan. Nurtured by Zen, these ink arts flowered in the mid 1400s like an exotic blossom in its adopted country. The men of Zen were particularly partial to *sumi-e* and *shodō*, for through these media they were "able to suggest the richness of the visible world without actually copying it."[13] Within the strict limitations imposed by ink, paper, and brush, the artist did not try to depict nature so much as to suggest the vitality behind it. "Each stroke has its moment of life and death; all together assist to interpret an idea, which is life within life."[14]

The ink arts, particularly calligraphy, can be thought of as a type of active meditation, a creative act of centering. Each embodies the spirit of Zen—one activity covering all activity. Each is an outward expression of an internal form, an expression of *musō*, which has been called the "unchanging formlessness behind all phenomena."[15]

The paper is the ground of activity acted upon by the energy of the painter as he abandons himself to the brush. The blank paper is *yin*, the receiver; the brush is *yang*, the agent. As the brush is wielded and the ink stroked and splashed upon the sheet, a dynamic balance between dark and light emerges. Neither white paper nor black stroke is static—the empty spaces push against the strokes as much as the strokes push into the blank spaces. Since there is more yin space than yang strokes, more paper left empty than stained, there is dynamic balance. The one completes the other. What is not *expressed* in ink is *suggested* in blank paper. "Brush absent, idea present," say the Chinese.[16]

The calligrapher is not restricted to established forms or styles of characters, nor even to making them clearly recognizable. What restricts are the tools and materials themselves. Paper, as Shiryū Morita suggested to me in interview, is a kind of interface between the truth as the artist realizes it and his discovery of it in artistic expression. When the artist strives for the highest through his art, he eventually comes to see paper as a limiting factor in his realization of his Buddha nature. Paper, like the brush and the ink, is both the obstacle to getting there as well as the way there. At the moment of emptiness, when he abandons himself to his materials and the strokes flow onto the paper in dynamic obedience to the Law—then the brush man expresses a freedom that does not come from his own clear spirit so much as from the act of abandonment itself.

### *Paper and the Artist*

THIS section consists of interviews with some of the leading Japanese artists and artist-craftsmen. All but one interview are written from notes taken as I talked with the artists themselves. I truly feel fortunate and honored to have met these men; I only regret being unable to meet the late Shikō Munakata, for his work I admire most.

Washi is central to the art of all these men. The quality of their work depends to a large extent on its beauty, strength, and durability. These artists have experienced washi intimately, and use it extensively in their creations. All of them possess a strong affection and high respect for Japanese handmade papers. But they speak for themselves.

### *Shiryū Morita, Calligrapher*

Born in 1912 in Hyōgo Prefecture, Morita received early recognition by winning the Ministry of Education Prize at the Japanese-Chinese Calligraphy Exhibition at the age of twenty-seven. In 1951 Morita founded and published *Bokubi*, a magazine of calligraphy and ink painting arts. Over the years he has exhibited in many international shows of calligraphy and abstract art. In 1963 he traveled throughout Europe by invitation of the West German government. A soft-spoken and mellow person who devotes some time to teaching advanced students, Morita is active in lay Zen circles. He resides in Kyoto.

"Man is dependent on his environment. But man must break his environmental limits to find freedom. A Zen man does this by way of Zen. I, through the brush. I want to find myself in my brush, to be conscious of my self within the brush. Completely, I seek free movement of the brush. A brush man becomes alive when he extinguishes himself and becomes the brush.

"My brush is my tool, my way of getting there. 'There' means a feeling of *mu*, of nothingness. To me the brush is at once my means of getting there, and also my obstacle to getting there. To overcome this obstacle, I must *become* the brush.

"Paper, too, is part of the environment, or the meeting place between myself and the environment, and so also a limiting factor in attaining freedom. The forming of the characters is frustrated by the paper, so I must overcome its limitations. In the process of this struggle, I can find *mu* in the Zen sense, and I can liberate myself. If I could paint without paper, and without a brush too, it would be perfect. Then I wouldn't even need *shodō*.

"Concerning paper, there are four important factors: (1) shape, (2) surface, (3) color, and (4) permanence. Paper plays a kind of feminine role in *shodō*. It is the passive part, the receiver. The strokes of ink enhance the whiteness of the paper, and the paper enhances the blackness of the ink. Both must be used to their fullest without subverting the other. Because calligraphy is dependent upon paper, I am dependent upon it also. In trying to overcome this obstacle, "I" must transcend it.

"The *gasenshi* I use is from China. I feel friendlier toward it because *shodō* came from China. I like the kind with some roughness. Western chemical pulp paper is no good, because it does not absorb the ink. The softness of the brush doesn't work on it. Also, I don't size my paper.

"It is impossible for people to escape ugly things. So if people want to be free or at peace, it is themselves that they must change. They probably need some tool for doing this. For some people it's Zen. For me it's *shodō*. Through *shodō* I try to feel *mu*.

"What is painting? It is like someone melting himself into the shadows of waves after a storm. Then the three bodies become one—the viewer, the shadow of the wave, and the sea."

*Keisuke Serizawa, Stencil Artist and Dyer*

Serizawa was born in 1895 in Shizuoka. He graduated from a Tokyo industrial arts school, studying graphic design. After graduation, he says, he felt like doing something special or unusual. For some time he did batik dyeing, but gradually came to feel that this was not true dyeing. While designing and studying dyeing at a relative's house, he met Sōetsu Yanagi, which changed the course of his life. Yanagi, just back from the Ryūkyū Islands and enchanted with *bingata* dyeing there, introduced Serizawa to Okinawan textiles. Serizawa fell in love with the bright colors and graceful designs. Around 1940 he went to Okinawa to learn *bingata*. From then on his dyeing followed *bingata* and folk art traditions. His use of paper as a dyeing surface came about during the war, when any piece of cloth was at a premium. The beautiful stencil-dyed washi endured, to the surprise of many.

Today Serizawa is a "Living National Treasure." His Tokyo workshop with its many young apprentices turns out much of the lovely hand-dyed papers, cards, and calendars sold in folk art shops and department stores throughout the country. A few of his former students are Seikichirō Gotō, Sadao Watanabe, and Yoshitoshi Mori.

"A most wonderful thing is a plain white sheet of *kōzo* paper!

"I myself use *kōzo* paper made in the town of Yatsuo, Toyama Prefecture. I know a papermaker there who selects paper for me and also makes it for me to suit my needs. There are many kinds of paper I could use, but my main requirement is for pure-fibered *kōzo* paper—paper containing no chemical pulp. This is because in the stencil process the paper sits in water for some time. *Kōzo* paper is strong in water. Paper machine-made of chemical pulp can endure water to a certain degree, but when you pull it out, it falls to pieces. *Mitsumata* paper, too, is generally too weak for this.

"My teacher, Sōetsu Yanagi, was a truly great man. He helped support many of us while he encouraged us to add new life to the great old traditions of folk art. Yanagi spent a lot of energy encouraging papermaking in villages all over the country, including the town of Yatsuo."

*Sadao Watanabe, Stencil Print Artist*

Watanabe was born in 1913 in Tokyo. He studied under Sōetsu Yanagi and Keisuke Serizawa. Watanabe has taken the traditional stencil dyeing methods and, working on various types of washi, has created beautiful, award-winning prints. In 1958 he won First Prize at a New York showing of contemporary Japanese prints. As a Christian, Watanabe is fascinated with Biblical themes. His iconlike style, which is both modern and timelessly traditional, has an immediate appeal to both East and West.

"I am much obliged to washi because it is so vital to my lifework, my prints. Even my wife has deep connections with paper—she is from a Yoshino papermaking family, and once made paper herself.

"Mainly I use *kōzo* paper from Toyama or the Sendai area for its quality and strength. *Momigami*, which I use for many of my prints, I order from Sendai. Most *mitsumata* paper is strong, but *kōzo* paper is stronger and works best for me. Whereas *mitsumata* is lovely, feminine, pliant, and soft, *kōzo* is tough and masculine, and that is the feeling I prefer for my prints.

"Several years ago I was invited to teach in a small college in Oregon. We devised ways to make resist paste out of available materials, namely rice flour and rice polish. We had to work very quickly with it. None of my American friends or students had much knowledge about Japanese papers, but all of them loved to receive gifts of *shōji* paper.

"I love washi. It is truly beautiful. To paint on washi with *sumi* ink by brush complements its character; it is hardly suitable for today's ballpoint pen. Machine paper made of pulp has no warmth to it. In washi, the human warmth of the maker shines through. People are just beginning to appreciate this warmth and seek it out. Many Western artists use machine-made papers for their prints, but I would never think of it."

*Hodaka and Chizuko Yoshida, Woodcut and Mixed Media Print Artists*

> Born in Tokyo in 1926 and 1924, this husband-wife team are well known and highly respected internationally for their prints, particularly their woodcuts. They have traveled and lectured through North America, Mexico, Europe, the Arab countries, and Asia. In Japan and abroad their work frequently appears in international print exhibitions. Here, the husband, Hodaka, speaks for the two of them.

"We test our paper by one of two ways, by tearing it or by using it. The stronger the paper is, the better. We have found sometimes, to our surprise, that old paper, say that made around 1930, is often better than the newer papers we purchase now.

"We like to use *hōsho* paper for our woodcuts. It is actually, we believe, stronger wet than dry. Sometimes we use *torinoko* too. All our paper comes sized. Nowadays some papermakers are beginning to size their own paper, but in general special people working for the paper merchants size it. They use *nikawa*, the glue of deer antler and bone. With such sizing, there is a problem of mold and spots growing on the paper's surface during the hot and humid summers here. I don't think that's a problem with unsized papers.

"We get our washi through a paper dealer in Tokyo, selecting from his sample book. We used to favor the *hōsho* made by Ichibei Iwano in Fukui Prefecture, but once he became a "Living National Treasure" we couldn't afford it. Instead we buy the *hōsho* of Mr. Yamaguchi in the same town.

"Good paper stretches little. We dampen our sheets with a brush and water before we print on them, so that they are already stretched a little. For etchings we use a very white Italian paper that is close pored. So except for practice sheets and the Italian paper, we use only washi."

*Shikō Munakata, Woodcut Artist*

Born in 1903 in Aomori Prefecture, in the northernmost corner of Japan's main island, Munakata appears to have been influenced early in life by the bold, dramatic and sometimes sensuous paintings characterizing the kites, banners, and illuminated floats for which that prefecture is famous. At the age of twenty-five he studied under the great woodcut artist Un'ichi Hiratsuka. In 1936 Munakata met Sōetsu Yanagi, Kanjirō Kawai, and Shōji Hamada, all of whom influenced Munakata with their reverence for folk art. Entries and awards in numerous national and international exhibitions followed; Munakata received the national Cultural Award in the sixties. Although his work is not limited to woodblock prints, they were his favorite medium. He died in 1975.

Munakata loved to depict Buddhist figures and themes, which he produced in astounding numbers with directness, vigor, deep emotion, and sensuality.

The following remarks are not first-hand. Since Munakata was unavailable to interview, the statements that follow were culled carefully from his writings and quotations.

"To me the woodcut print is mysterious and beautiful because it is universal—there is no self to it. Rather, self and world become one as I throw myself completely into the making of it. No trace of self remains. And so in this sense I am not responsible for my work. It springs not from ideas or intellectualization, but from infinity. It is brought forth from an integrity that cannot be differentiated by either 'this' or 'that.'

"It is impossible for a woodcut print to be an ugly work—it is truth even when it makes mistakes. Man creates ideas of beauty and ugliness; neither the wood nor the woodcut create them. The world of the woodcut is vast and boundless, and that is why it appeals to me.

"Paper, wood and knife—these are the essentials, nothing more. And so, working quickly, purity and vigor is maintained in the work. A sheet of plain white washi is perfection itself, and so is the block of wood. If I cannot create on that empty space something more beautiful than those plain surfaces, I have done nothing at all. A woodcut print springs from reality with great intensity and power; I hope that woodcuts will show us the truth of infinite beauty."

*Hiroshi Haneishi, Nihonga Painter*

Haneishi is a *Nihonga* painter and member of the Japan Art Academy. (*Nihonga* is the traditional Japanese manner of painting and drawing.) Haneishi's work involves repairing and copying old art works as well as creating original paintings in the style of numerous old schools. The dragons he painted inside a certain Nikkō temple are National Treasures. Haneishi was kind enough to show me one of his most prized possessions, a piece of the *dhāraṇī*, one of the oldest extant examples of printing on paper, dating from 770 A.D. The hemp paper, backed for protection, was delicate, slightly brittle, and a tender brown color. The box with the small hand scroll also contained a tiny piece of expensive incense.

"I am very strict in my selection of papers because they are so vital to my work. Did you know that good paper lasts longer than silk? It does. Each artist must select his paper according to his own taste. Because I am known, I have no difficulty obtaining good papers, fortunately.

"Like most *Nihonga* artists I use *mashi* (hemp paper). There are many kinds of *mashi* even today, differing in character depending on where it is made. It is possible to find huge sheets of it. Sometimes I use good quality *hōsho* as well. The *torinoko* paper as made of old, with *gampi* fiber, is no longer available. This was once called *kami no kamisama*, "the god of papers." It had an incredible refinement. The *torinoko* made in the past hundred years tends to crack with age. It is so delicate that it is easily torn and it loses its luster if worked over. *Mashi*, on the other hand, is soft and pliant; it maintains its fine quality even if worked over repeatedly.

"In the old days, warriors sent messages to the enemy by wrapping letters written on *gampi* paper around arrows and shooting them into the opposing camp. *Gampi* paper then was used for everything from hair ties to sacred letters. From time to time I can obtain some old paper, which I love to work on. When you paint on old paper, you don't have to size it. Its age prevents the ink from bleeding excessively. In general, paper without sizing has a more tasteful feeling. *Gasenshi* and other unsized papers, by the way, are excellent for making rubbings.

"The colors I use are the very same ones used in ancient times. Nothing has changed. The pigments are derived from natural minerals. I mix the powdered color with a boiled glue binder made from fish bones or animal hooves.

"In good *Nihonga* one must be careful not to draw or paint too much. That is important. Some of the paper must be quite empty. Let the paper live!

"You ask me if I have some special feeling toward old washi or works of art done on old paper. If I see really old paper of excellent quality, I want to paint on it. And when I see a very beautiful ancient scroll or painting on *fusuma*, I am speechless. And yet sometimes I have the feeling that even a new *Nihonga* painting immediately has an ancientness, an old feeling.

"How old am I? (wave of hand) Young! If I think of myself as old, my painting will be old!"

NOTES

half-title: Sōetsu Yanagi, "Oshie," excerpted in *Ginka* 9 (Spring, 1972): 36–39. (trans. by Hitoshi Hosokawa)

1. Sōetsu Yanagi, *The Unknown Craftsman*, trans. Bernard Leach (Tokyo: Kodansha International, 1972).

2. Toshio Ikeda, ed., *Izumo no kami* (Iwasaka, Shimane Pref.: Eishirō Abe private pub., 1952). (trans. by Hitoshi Hosokawa)

3. Yanagi, "Oshie," *op. cit.* (trans. by Hitoshi Hosokawa)

4. Yanagi, *Unknown Craftsman, op. cit.*

5. Tetsurō Watsuji, *Climate and Culture*, trans. Geoffrey Bownas (Tokyo: Hokuseidō,

6. Yanagi, *Unknown Craftsman, op. cit.*

7. Kakuzō Okakura, *The Ideals of the East* (Tokyo and Rutland, Vt.: Tuttle, 1970).

8. Yanagi, *Unknown Craftsman, op. cit.*

9. Shin'ichi Hisamatsu, *Zen and the Fine Arts*, trans. Gishin Tokiwa (Tokyo: Kodansha International, 1971).

10. Issōtei Nishikawa, *Floral Art of Japan* (Tokyo: Japan Travel Bureau, 1949).

11. Tokutarō Hanada and Kiyofusa Narita, *Kami-suki uta*, 2 vols. (Tokyo: Seishi Kinenkan, 1951). (trans. by Yasuo Sasaki)

12. Sōetsu Yanagi, "Washi no bi," *Ginka* 9 (Spring, 1972): 32–36. (trans. by Louise Picon Shimizu)

13. Seiroku Noma, *The Arts of Japan*, Vol. I, trans. John Rosenfield (Tokyo: Kodansha International, 1966).

14. Okakura, *op. cit.*, p. 180.

15. Yanagi, *Unknown Craftsman*, *op. cit.*, p. 121.

16. Al Chuang Huang, *Embrace Tiger, Return to Mountain* (Moab, Utah: Real People Press, 1973), p. 133.

*Waters of the valley stream are drawn*
*to each house;*
*Papermaking maidens singing, singing songs.*

# The Japanese Papers

MORE KINDS OF HANDMADE paper are made in Japan than in any other country in the world. Many, many more types of washi are produced than are mentioned in this chapter. The papers discussed here are those I consider most worthy of mention, a fairly representative selection of names that one comes across most frequently when dealing with washi. Since new handmade papers are constantly being developed, and since old types of papers just as quickly die out, it is doubtful whether anyone has ever been able to keep an up-to-date catalogue of them all.

In deciding upon a chapter such as this, I often asked myself whether or not it would be really of interest or use to many readers. My conclusion was that if it was helpful to even a few, it would be worth researching and writing. The first part of the chapter presents more or less "plain" papers, including both historically famous and more recently developed types, many of which descend from the historical papers. Their origins, characteristics, the changes they have undergone, and, in many cases, their demise are traced. The remainder of the chapter is devoted to "special papers," that is, papers dyed, treated, watermarked, imbedded, or the like, and the elaborately designed *sukimoyō-gami*, which so amaze Westerners—all papers devised to meet an old and constant demand for novelty and variety in washi.

There are so many varieties of washi that I cannot pretend to be well acquainted with them all. Not only are there hundreds of different types, many with colloquial names, but the nomenclature is compounded by a profusion of descriptive, specific names that designate such things as where a paper is made, its materials, its color, its pattern, its use, etc., as well as any combination of these. For example, *shiro-mashi* designates a white (*shiro*), hemp (*ma*) paper, so named to distinguish it from colored or dark, unbleached papers. *Suruga-banshi* describes a *hanshi*-type paper once made in the Suruga area. *Mizutama-torinoko* is a *torinoko* paper overlaid with a dot or circle (*mizutama*) pattern. Moreover, *Echizen koban bijutsu torinoko kusairi* is literally a "small-sized sheet of artistic *torinoko* paper from Echizen imbedded with pieces of grass." One begins to see the problem.

The classification of the papers included here is necessarily subjective. The larger categories are merely a convenience; in reality, as in life, divisions are fluid and overlap. *Yakutai-shi*, for example, is a historically important paper and so appears in the first section along with the common papers *shibu-gami* and *fuda-gami*—yet actually they are all also "treated" papers. Also, papers named after their place of origin are capitalized, although some other writers do not do so. It matters little, but it may help identify readily a paper whose name is unfamiliar; if the name is capitalized, it would be a fair guess that it designates where the paper is or was once made and thus, for those familiar with an area's reputation, also possibly give a clue as to the paper's quality.

I should mention here also that unfortunately the names given to papers sold by Western retailers more often than not are completely different from their true Japanese names, no matter how authentic these names may sound. Not always, but more frequently than not. For instance, what is sometimes in the United States called "taiten" is really an *unryūshi*; "wakasa" is a variety of *chiri-gami*, possibly from the Wakasa Bay area; "sumi-e," a *gasenshi* or even a *wara-banshi* (often used for *sumi-e* painting practice); and what is called "hosho" is usually not a *hōsho* at all but a bastardized variety that contains chemical pulp. The creamy woodblock paper one retailer calls "Abe," on the other hand, looks like a fine paper and could well be the product of Eishirō Abe's mill. Why names of papers are changed when they are imported by other countries I am not sure. Aside from there being special codes of the wholesalers, I suspect that it is at least in part to keep the uneducated purchaser ignorant of the true source and quality of the paper and to enable the dealer to substitute more easily a similar paper for one that has become too expensive or is no longer made. In effect, this custom just adds to the general confusion of those trying to use and identify papers. For some hints on how to judge Japanese papers, consult the Evaluating Washi section at the end of this chapter.

In theory, the papers that follow might be further analyzed and categorized into smaller groups. In practice, all categories of papers overlap at some point or other. To aid in giving a more precise picture of these papers, the following descriptive terms are used:

Common—commonly made
Local—production limited to one area or district only
Special—occasional or highly limited production
Vanishing—manufacture may be dying out
Defunct
Quality

*Plain Papers*

*azabu-shi* 麻布紙 (local; vanishing)

This pure *kōzo* paper, made since the early seventeenth century, is still made in Kaminoyama, Yamagata Prefecture. *Azabu-shi* is a variety of *urushi-koshi*, extremely thin, stiff, tough, and very porous. It resembles *Yoshino-gami* and is used for the same purposes, filtering lacquer and high-quality chemical paints; sometimes the two papers are used jointly. A piece of silk gauze that has been treated with persimmon tannin or that has been lacquered is used in its scooping, preventing the fine fibers from escaping through the screen. The sheets are couched directly onto the drying boards as soon as they are scooped.

*cha-gami* 茶紙 (vanishing or defunct)

This *mitsumata* paper, literally "tea paper," is used for sealing the wood boxes used to store and transport tea leaves. The town of Kiyosawa in Shizuoka Prefecture was particularly well known at one time for producing *cha-gami*, but it is very doubtful any is made there now. It was natural that this kind of paper once flourished here, for it is the center of Japan's tea industry and the country where *mitsumata* abounds in both its wild and cultivated states.

*chiri-gami* 塵紙 (common)

*Chiri-gami* means literally "waste paper" or "dregs paper." As long as paper has been made from bark, *chiri-gami* has been made, and every papermaker once manufactured it. Traditionally *chiri-gami* was made in whole or part from the outer dark bark and fiber residue cleaned off of the white bark fiber. So while it is not really recycled paper, it is an economical use of the outer bark normally considered unsuitable for fine quality paper and that otherwise would be discarded.

It is said that a papermaker's *chiri-gami* is a good measure of his skill and integrity as a paper craftsman, for if he conscientiously cleans his fiber, the residue, or *chiri*, will contain many good strong fibers along with the rejected ones. The presence of these bits of *chiri* in itself is proof that the papermaker does most of the work by hand, not machine. Today some *chiri-gami* is made by boiling the black, uncleaned bark in an alkali and using that substance for paper, but this is considered a lazy papermaker's *chiri-gami*. Some sophisticated papermakers such as Eishirō Abe make a very pleasing paper by adding flakes of black bark to the vat solution, calling the paper *chiri-iri*, or *chiri*-"added."

*Chiri-gami*, which was until recently used as toilet paper and handkerchiefs, is appreciated highly in the West for its natural beigy tones and interesting texture formed by the bits of bark and fiber running through it. Although not as strong as, say, the finest *gampi-shi* or *hōsho*, it is nevertheless usually of good quality, providing it is made by the best traditional methods. Depending on its strength and appearance, it is used for bags or wrapping or for artistic purposes such as art printing and as endpapers in books.

*chōchin-gami* 提灯紙 (common)
*Chōchin* means lantern, and *chōchin-gami* is therefore any paper used to cover lanterns. Usually this means a soft, thin *kōzo* paper that can withstand the elements and allow light to pass through it. Sometimes, however, *mitsumata* paper is used. The Mino area in Gifu Prefecture is famous for its *chōchin-gami*. It is also made in Fukuoka Prefecture in Kyushu and in the Tōhoku area, in which the annual Nebuta Festival is famous for immense paper lantern floats paraded through the streets at night, painted with fantastic scenes from history and legend and constructed in the shapes of figures and animals.

*fuda-gami* 札紙 or *shibu-gami* 渋紙 (common)
*Fuda* means label or tag. This paper is coated with *shibu* (persimmon tannin) to make it waterproof. Traditionally it was used for price tags on dry goods, or the thick variety was used as an umbrella paper. Today it is most often used to label cloth that is to be dyed. Also, small sheets of it are sometimes made for the repair of holes in very old paper *shōji*. *Fuda-gami* is still made in Kurodani, Kyoto Prefecture (see Shibu-Treated Papers).

*gampi-shi* 雁皮 (special; quality)
*Gampi-shi* is the generic term for all papers made of pure *gampi* fiber. It was one of the first papers developed in Japan, appearing in the Nara period (710–794) under the name *hi-shi*, which it is occasionally still called. Reputedly *gampi* papers are the hardest to make. Generally a piece of silk gauze is used over the screen when scooping to prevent the loss of these delicate fibers; therefore, *gampi-shi* does not bear the chain-lines characteristic of other papers formed directly upon the screens.

*Gampi-shi* is glossy, beige in tone, small pored, very delicate, and slick to the touch. It is highly translucent. The fibers are long, thin, and very strong. *Gampi-shi* is pliable and does not crease easily, although it has a natural wrinkle, especially along the edges, as if the center of the sheet had dried much more slowly than the outside edges. It is made in several thicknesses, from the very thinnest translucent sheets to thick ones with curled edges.

*Gampi-shi* is not particularly suited for printing because of its tendency to shrink. Rather, it is best suited for writing and calligraphy when great absorption of ink is not desirable. It is also well suited for fans; as protective covers around and sheets within books; as drafting and tracing paper; for wrapping precious objects; for square poetry paper (*shikishi*); and especially for hammering out sheets of gold leaf.

Because the *gampi* plant has not lent itself to cultivation in quantity, the plant is becoming increasingly rare, and very little *gampi-shi* is made today. The Naruko family in the village of Ōmi-Kiryū, Shiga Prefecture, carries on the making of *gampi-shi* according to the finest old traditional methods. This workshop is famous for *Ōmi torinoko*, an Intangible Cultural Property of Shiga Prefecture, which is used to repair old cultural treasures of paper. "Living National Treasure" Eishirō Abe is another maker of fine *gampi-shi*, having received his title in the main for his *gampi* papers.

*gasenshi* 画仙紙 (common)
*Gasenshi* is a type of traditional Chinese drawing paper primarily used in *sumi-e* and calligraphy. *Gasenshi* today is chiefly made of *mitsumata* combined with a smaller amount of other fiber such as bamboo, straw, cotton, recycled paper, and often other fibers as well. The addition of bamboo or straw make *gasenshi* quite absorbent and soft.

A man by the name of Kiyokatsu Shimizu in Misumi, Shimane Prefecture, makes a fine *gasenshi*. It is also made in other parts of west central Honshu, notably in Aoya-chō, Tottori Prefecture; and in other areas of western and southern Japan.

*geijutsu-shi* 藝術紙 (local; special; quality)
*Geijutsu-shi* translates as "art paper." This rather expensive paper is made of *mitsumata* and is very thin, unbleached, and brownish in tone. For the most part, it is a high-class wrapping paper. *Geijutsu-shi* is brushed onto wood boards in special rooms where steam comes up through the floors. This gentle drying treatment helps it retain its delicacy and pliancy. It is used for packing gold leaf, pottery, lacquerware, paintings, and other delicate items. *Geijutsu-shi* is produced primarily in the town of Ikazaki in Shikoku.

*hakuai-shi* 箔合紙 (special; quality)
This is the special, pure *mitsumata* paper used to package metal foils (mostly gold). Such foils are sold in packs, usually in multiples of ten sheets, and each sheet is sandwiched between *hakuai-shi*. This paper is oiled so that one surface is shiny and the other matt. The purpose of this paper, other than the obvious one of protecting the delicate foil, is to provide a thin, removable backing to which the foil will adhere without slip-

ping while a craftsman is working with it. The oiled surface of *hakuai-shi* loses its adhesive quality in time, but this can be restored with a light touch of rapeseed oil.

*haku-uchi-shi* 箔打紙 (special; quality)
This term means "foil-beating paper," and it is used, as one might expect, in the manufacture of fine metal foil sheets. The quality and thinness of the foil is a direct result of the quality of the paper used. This is a pure *gampi* paper that either contains a little clay of mainly silisic acid and alum or is coated with persimmon tannin or bears both clay and tannin; the clay and tannin fill in the pores, making the sheets hard, smooth, and slippery so that when the gold is hammered between the sheets it will adhere to them. Each sheet is thinly coated with metallic dust in layers of three hundred sheets at a time, then hammered together, so the paper must be fine and strong. *Haku-uchi* is very thin and an off-white color; after beating it becomes translucent, absorbing oil well, and is then used as a cosmetic paper. This unusual paper is still made as it has been for centuries, near prosperous centers that can support gold lacquerwork and Buddhist art, two such producers being the papermakers in Najio, Hyōgo Prefecture (see *maniai-shi*), and the Chika (widow of Saichirō) Naruko workshop in Ōmi-Kiryū, Shiga Prefecture.

*hanshi* 半紙 (common)
The most common paper made from the middle ages until recent years was *hanshi*. *Hanshi* means half (-sized) paper, being half the width of the old standard sheet of *Sugihara-shi*. This half-sized sheet, which measured about 25 × 35 centimeters, proved to be useful and practical and, because it was also durable, thin, light and relatively inexpensive, became popular with everyone, rich and poor alike, as a paper for daily use.

Although *hanshi* has always been a paper of a thousand uses, of old it was mainly a paper for books, especially account books. Made of pure *kōzo*—with the exception of *Suruga-bashi*, which was made of *mitsumata*—*hanshi* was famous for its water-resistant qualities. In case of fire, account books made of it were protected by being thrown into the nearest well.

Every province produced its own *hanshi*, and it became practice for the finest of these products to bear in some form the name of their area of manufacture. Thus, some high quality *hanshi* papers such as *Suruga-banshi*, *Sekishū-banshi*, *Tokuji-banshi*, and *Yanagawa-banshi* (from Chikugo) became famous throughout the country.

Today's *hanshi* is either pure *kōzo* or, more commonly, *kōzo* with some pulp added.

*hiki-awase* 引合 (defunct)
This old *kōzo* paper is believed to have been another name for *danshi* (see p. 206).

*Hiki-awase* is the inner lining of a medieval Japanese warrior's breastplate, which consisted of layers of these tough sheets of paper.

*hishi* 斐紙 (defunct)
*Hishi* is an old name for *gampi-shi* and is related to the original *torinoko* paper. The term *gampi-shi* has been in use only since the end of the Edo period.

*Hodomura (-shi)* 程村紙 or *Karasuyama-washi* 烏山和紙 (local)
*Hodomura* was well known throughout the Edo period as a *kōzo* paper carefully made of the finest fibers and manufactured in what is now Tochigi and Ibaraki prefectures. It is still made today in the town of Karasuyama, Tochigi Prefecture, although it is not of the quality it used to be. *Hodomura* is creamy colored, strong, and porous. In Japan it is used for printing books, and in the West for art printing.

*hon Takakuma* 本高熊 or *Takaguma*; *Takaguma-gami* (local; vanishing)
This *kōzo* paper originated in some small mountain villages in Toyama Prefecture, where the fibers were laid upon the snow to whiten. *Hon Takakuma* was often waterproofed and used primarily as an umbrella paper; but it also was used at one time in history for the bags containing powdered drugs and medicines for which Toyama became famous. Today it is still made in the town of Yatsuo, where it is dyed with stencils, producing many of the popular and enchanting folk art design papers seen marketed today. Of this paper, Bunshō Jugaku wrote in his *Paper-Making by Hand in Japan*: ". . . *hon Takakuma* is unpretentiously white, noble, crisp, charming, and yet sedate; both to touch and sight, it inspires affectionate reverence, if not awe."[1]

*hōsho* 奉書 (common to special; quality)
True *hōsho* has always been a superior type of pure *kōzo* paper, resembling the traditional *Sugihara-shi*, of which it is perhaps an offshoot, but of higher quality. *Hōsho* first appeared in the sixteenth century in Echizen (Fukui Prefecture). *Echizen hōsho* was a favorite of the samurai from the late Muromachi period (1333–1573) and of the court nobles beginning in the Edo period (1615–1868), who used it as ceremonial wrapping and writing paper. Gradually many villages began manufacturing *hōsho*, especially in central Japan and Shikoku, and eventually the quality of this grand paper deteriorated.

Good *hōsho* is thick, strong, fluffy, creamy white, and absorbent—the finest paper available for Japanese-style woodblock prints. It is made to withstand numerous rubbings and it does not shrink, expand, or tear easily. Although very fine *hōsho* was once made in Iyo (Ehime Prefecture), now chemical pulp and starch are used there in its production.

This inferior imitation of true *hōsho* is often referred to as either *hosho* or *hosho-shi*, or abroad as Iyo. It may be considered suitable for inexpensive art printing, where paper strength and endurance are not of primary importance. This paper, however, should not be confused with pure *kōzo* fiber *hōsho*, sometimes referred to as *kizuki hōsho* ("purely made *hōsho*"), which is greatly superior.

*Hōsho* in its best traditional state is still produced, as far as I know, only in the village of Ōtaki, Fukui Prefecture, in the workshop of the late "Living National Treasure" Ichibei Iwano and his family; and possibly by Tadao Endō of Shiroishi.

*Hosokawa-shi* 細川紙 (local; quality)
A *kōzo* paper similar to *Sugihara-shi*, *Hosokawa-shi* has an interesting history. During the Edo period, a branch of the ruling Tokugawa family controlled the Kii area and brought its paper to Edo. A kind of *hōsho* made in the town of Hosokawa near Mt. Kōya and Mt. Takano was particularly favored, although rather expensive due to the great distance between the papermaking town and the markets. At some point the techniques of making *Hosokawa hōsho* were brought to the Musashino area, the plain stretching north and west of Edo. Musashino was already engaged in making *Musashi-shi*, a good-quality *kōzo* paper, which had direct Korean influences. Techniques of making these two papers were combined, improved somewhat, and *Hosokawa-shi* was born.

Some historians hold that the name *Hosokawa-shi* developed in order to hide its true place of manufacture. Evidently some powerful paper wholesalers in Edo banded together to corner the market on *Hosokawa-shi*; the fact that it was really produced so close to the Edo markets was a highly guarded secret. Its fine quality and close proximity to those capital markets made the production of *Hosokawa-shi* a flourishing industry, giving rise to the expression *Pikkari, sen ryō*—"A flash of sun, a thousand *ryō*." A bright day meant lots of crisply dried paper—and money (then *ryō*) in the papermakers' pockets. Those who made this paper probably gave it the name *Ogawa-shi*, from the papermaking town of Ogawa in Saitama Prefecture, the center of production. Today *Ogawa-shi* is the name for several papers evolved from the original *Hosokawa-shi* but having undergone modifications of mechanized manufacturing techniques—inferior offspring.

Historically *Hosokawa-shi* was a paper for government land records and other books of permanence. It was also waterproofed and used for umbrellas. Today Saitama Prefecture does its best to keep and preserve *Hosokawa-shi* in its original excellence, as well as to preserve old papermaking tools. These efforts have been rewarded, since *Hosokawa-shi* has been named a *mukei bunkazai*, an Intangible Cultural Property.

*Iwami-washi* 岩見和紙
A general term for various papers made in the Iwami area of Shimane Prefecture, such as *Sekishū-banshi*.

*Izumo mingei-shi* 出雲民藝紙 (local; common; quality)
Literally, *mingei-shi* means "folk art paper." *Izumo mingei-shi* specifically refers to some fifty or more varieties of paper made of *gampi*, *kōzo*, and *mitsumata* and their combinations produced at Eishirō Abe's mill in Iwasaka, Yatsuka-gun, Shimane Prefecture. Izumo is the ancient name for that area. A product of Mr. Abe's elevated taste and aesthetic awareness, these well-made and beautiful papers retain a certain folkcraft character.

Izumo paper has an old history, originating in the northern Shimane area in the eighth century. Eventually the craft devolved into the making of poor quality straw-base paper used as calligraphy practice sheets.

Abe revived the art of making fine Izumo paper in the 1950s, experimenting to develop new and interesting papers in the folk art tradition. *Izumo mingei-shi* includes fine cream-colored *kōzo* papers, *chiri-iri*, dyed papers, *unryūshi*, *mitsumata*, and *kōzo* fiber papers, *gampi* papers, and numerous other simple and designed papers, mainly for artistic and decorative use.

*Izumo-shi* 出雲紙
Originally meaning a kind of *wara-banshi* of rice straw made around Izumo in Shimane, *Izumo-shi* now generally refers to the folk art papers of Eishirō Abe (see above).

*kairyō-shi* 改良紙 (common)
A relatively new paper, *kairyō-shi* translates as "improved paper," which really means that a lot of machinery is used in its production and therefore it lacks the natural qualities of most handmade washi. This thin *mitsumata* paper is used as calligraphy paper as well as copy and typing paper. It can take considerable handling and folding. *Kairyō-shi* is still made in Kawanoe, Ehime Prefecture, Shikoku. At least until recent times it was also made in Kōchi and Tottori prefectures, but I am uncertain whether or not it is still produced there today.

*kaishi* 懐紙
*Kaishi* is not a particular paper, but the term for a number of high-quality papers associated with the Tea ceremony. *Kaishi* literally means "pocket paper"; its present form is the descendant of the widely used utility paper tucked under the top edge of the *obi*

or in the front fold of the kimonos of either sex, the only "pocket" in traditional Japanese clothing. Today *kaishi* is used to cleanse the teabowl where one's lips have touched it. This paper also serves as a kind of plate for the cakes served in the ceremony. *Kaishi* is generally *danshi*, *hōsho*, *hanshi*, or *hanagami*.

*kasa-gami* 傘紙 (common to special)
*Kasa-gami* means "umbrella paper" and refers to any paper used in the manufacture of umbrellas. *Kasa-gami* must be made with great care and of the best materials, according to Bunshō Jugaku. Usually they are rather thick, pure *kōzo* papers, which have been coated with oil to waterproof them and thereby also referred to as *abura-gami* ("oil paper"). Often they are dyed in beautiful colors and color combinations. About twelve sheets of 45 × 30 centimeter paper make up one umbrella. Umbrella papers included here are *hon Takakuma*, *Morishita*, *senka*, and *uda-gami*.

*kikyū-genshi* 気球滅資紙 (defunct)
Literally, "balloon paper." A type of multipurpose paper called *genshi*, this strong but thin *kōzo* paper was once used in making paper balloons, the standard-sized sheets pasted together to form enormous ones. It is quite possible that *kikyū-genshi* was used along with other types of paper in making the notorious if abortive paper bomb balloons (see p. 186).

*kinpaku-gami* 金箔紙 (special; quality)
*Kinpaku-gami* means "gold-leaf paper," and is a form of *haku-uchi* (q.v.).

*kōzo-gami* (*-shi*) 楮紙 (common)
This term includes all papers made of one hundred percent *kōzo* (q.v., p. 77). *Kōzo-gami* was one of the first papers produced in Japan, its fiber originally used and developed for paper during the reign of Prince Shōtoku in the early seventh century.

Hundreds of varieties of paper made from *kōzo* exist in Japan today, ranging from rough, thick papers of brownish or creamy hues to thin, delicate, white ones. *Kōzo* is the most widely grown papermaking plant, wild and domesticated, and it naturally follows that more *kōzo* paper than any other is made by hand in Japan.

*Kōzo* fibers are long and sinewy compared to the other fibers used. *Kōzo-gami* is strong and durable, somewhat fluffy, absorbent, and has body. To some extent it is resistant to water, the paper having the ability to spring back into shape as it dries. Even thin, delicate-looking sheets such as those of *urushi-koshi* have a robust and warm quality to them. These qualities have prompted the Japanese for centuries to use *kōzo* paper for clothing, raincoats, umbrellas, bedding, lacquer filters, and *shōji* and *fusuma*. Its shrinkage being minimal, *kōzo-gami* is also excellent for woodblock, etching, and most every other type of printing. Moreover, the finest *kōzo-gami* reputedly gets whiter with age.

*Kōzo-gami* is also used in making kites, toys, lampshades, lanterns, stationery, book and endpapers—in fact in the manufacture of just about any paper product where beauty, strength, and durability is sought.

*Kōya-shi* 高野紙 (local; special; vanishing)
For hundreds of years *Kōya-uda* was a good, slightly coarse umbrella paper; now *Kōya-shi* is extremely close to extinction. Today only one woman and her daughter, who comes home in winter to scoop, continue the tradition in the town of Kudoyama, Wakayama Prefecture—a tradition that began many centuries ago on the slope of Mt. Kōya, one of the major temple complexes in Japan and the goal of pilgrimage.

*Kōya-shi* is a *kōzo* paper quite small in size, dark in tone, and bearing chain-lines from the miscanthus screen used in its scooping. Mostly wild *kōzo* fiber is used, and the women manufacture both a natural white paper and *chiri-gami*.

Quite unique methods are used in its making, old *tame-zuki* methods that have all but disappeared from Japanese papermaking. Thirteen molds, replete with screens, are used successively to scoop the *kōzo* solution, each mold then placed at an upright angle on a shelf or on the ground to drain while the next one goes through the vat solution. Then the molds are dipped again in succession until the right thickness is obtained. When sufficient water has drained through, the sheets are removed by hand from the molds, placed upon the drying boards, and brushed flat. No pressing takes place.

For years Kōya papermakers tried in vain to make a go of their dying paper industry. Now, ironically, the only *Kōya-shi* made today, by these two ladies, is eagerly bought up by an Osaka paper dealer and sold as the last of a vanishing type of washi.

*kyoku-shi* 局紙 (common)
*Kyoku-shi*, a type of *torinoko* paper, has been called "Japanese vellum." Thick and durable, it can withstand numerous foldings and rubbings. Two types exist, one of *kōzo* and a more prevalent type made of *mitsumata*. Traditionally the *tame-zuki* method is used in its production, particularly for the variety made of *mitsumata*. Calendering, pressing between rollers, gives it a smooth, glossy surface highly suitable for printing.

*Kyoku-shi* means "office" or "bureau" paper. It was first produced in the Meiji era for the printing of government bonds, stocks, securities, and other official papers. Then, as now, it was frequently watermarked to prevent counterfeiting or to lend the paper an official stamp.

The best quality *kyoku-shi* has for years been in great demand in Japan and abroad for deluxe editions of books and for official documents, not only for its superb receptivity to print and nonshrinking qualities, but also because it is repellent to silverfish. Today it is used in certificates, bonds and stocks, calling cards, art printing, and for many other uses where a smooth, high-quality, hard paper is essential. In recent years a paper similar to *kyoku-shi* but containing chemical pulp has appeared, bearing its name, but it has nothing of the quality of the real stuff.

*Kyoku-shi*, of both the *kōzo* and *mitsumata* varieties, is made for the most part in Fukui Prefecture. The type made in Ishikawa Prefecture, to my knowledge, contains chemical pulp.

*mafu*
Another name for *Yoshino-gami* (q.v.).

*maniai-shi* 間似合紙 (special; vanishing; quality)
*Maniai-shi* is an old variety of *gampi* paper containing clay. It evidently sprang up as an imitation of *Echizen torinoko*; but the pulverized rock peculiar to the area of Najio, Hyōgo Prefecture, which was first added to it in the middle ages, made it a distinctly new and different paper. The need for a more versatile type of *torinoko* paper arose during the Muromachi period (1333–1573), when the tea ceremony began to enjoy prestige. This need—for a new, larger paper for use in folding screens, *fusuma*, and hanging scrolls—was met by *maniai-shi*, whose name means "fitting" or "appropriate" paper, denoting its larger, more serviceable size. It was also called *doro-ire-torinoko*, *torinoko* with clay added.

In Najio today older men still make *maniai-shi* of one hundred percent *gampi* fiber. Moreover, they employ a unique method in manufacturing it, a method that lies somewhere between the *tame-zuki* (Western) method and *nagashi-zuki* (Japanese) method. This indicates an amazing adherence to ancient traditional methods, for *tame-zuki* (or *tame-nagashi-zuki*) is believed to be the very method introduced to Japan from China and Korea about fourteen hundred years ago. The papermaker sits cross-legged upon the ground and scoops paper from a low vat, "strokes" it quickly over the screen-mold, then allows the sheet to rest for some time and drain while still attached to the gauze-covered screen. The method is called *tame-nagashi* because it combines features of both methods; mucilage is used, and the screen is shaken somewhat more than in typical *tame-zuki* papermaking.

Besides the qualities imparted by the old techniques of scooping, *maniai-shi's* uniqueness comes from the addition of the special Najio clay, whose composition is still a matter of great secrecy. Only a small amount of the clay is used, enough to enhance the paper's quality but not enough to "load" it. Consequently, the paper has a somewhat greasy or soapy texture.

Most *maniai-shi* made in Najio today is used to preserve museum art works or to back *fusuma*, notably in old Kyoto temples. The papermakers at Najio believe that the old paper bearing antique art absorbs some of the clay from *maniai-shi*, thereby regaining lost moisture and suppleness. A very thin type of *maniai-shi*, called *usu maniai-shi*, is believed by many to have been the prototype for the Oxford University Press's "India paper" used in lightweight Bibles. The art of making *maniai-shi* is definitely near extinction, since no one under fifty-five makes it today in Najio and no new people have begun learning its techniques.

*mashi* 麻紙 (special; vanishing)
*Mashi* is paper made of hemp fiber. It is the oldest type of paper still being made. Most of the first paper introduced to Japan from Korea is thought to have been hemp paper, although it probably was a type manufactured from hemp rags rather than directly from hemp plant fiber. Eventually, as *gampi-shi* and *kōzo-shi* came to be more efficiently made, production of *mashi* greatly diminished.

Rather little *mashi* is made today, using methods and materials greatly improved through the centuries. Reputedly today's product is stronger than *torinoko*, besides being quite pliable. Lustrous and very white, it resembles a finely made, bright *kōzo* paper. *Mashi* is reputed to maintain its luster and high quality even when worked over extensively.

Some *mashi* is made in the Ōtaki village area of Fukui Prefecture and possibly still in Toyama and Niigata prefectures.

*Michinoku(ni)-gami* 陸奥紙 (defunct)
The remote northeastern region of Japan known as Tōhoku is said to be the birthplace of *Michinoku-gami*. This thick, water-resistant, tasteful, and very white paper appeared in the Kyoto area about 950–1000, where it quickly became a favorite of the court. A high-born woman who became inordinately fond of this unusual paper was Lady Sei Shōnagon. She even mentioned it in her *Pillow Book*, a bright and witty observation

of Heian court life. *Michinoku-gami* also appears in *The Tale of Genji*. Most paper scholars believe it to be the prototype for *danshi* (q.v.).

*Mino-gami* (*-shi*) 美濃紙 (local; common); *hon Mino-shi* (local; quality)
*Mino-gami* is a catch-all term for a number of plain *kōzo* papers originally produced in the province of Mino, Gifu Prefecture. These include such papers as *hon Mino-shi* (or *-gami*), *shoin-gami* (or *-shi*), *Morishita*, *tengujō*, *chōchin-gami*, and *bijutsu-shi*; many of these are discussed elsewhere in this chapter.

*Mino-gami* is reputed to be one of the oldest types of *kōzo* papers in Japan, appearing in the mid 700s according to some sources, although others put its first appearance as late as 1190. Papermaking in Mino took great leaps forward after the Ōnin War in 1467, and eventually so much *Mino-gami* was produced that, being rather inexpensive, it became a paper for daily use among the populace. In the Edo period, *Mino-gami* was used so commonly for books that its size determined books' standard size.

*Mino-gami* is still made according to fine traditional techniques by Kōzō Furuta in the village of Warabi, Gifu Prefecture. *Hon Mino-shi* is made of *kōzo* fiber plus a little chemical pulp for softness and absorbency. The *kōzo* fiber is partially hand-beaten with wooden mallets whose heads have been carved in a "chrysanthemum" pattern—a tool peculiar to this area. *Hon Mino-shi* has been designated a *juyō mukei bunkazai*, an Important Intangible Cultural Property.

When referring to *Mino-gami* today, most people mean a kind of *kōzo* paper used either for stationery and books or *shōji*, i.e., *shoin-gami* (q.v.).

*mitsumata-shi* 三椏紙 (common)
This term refers to all papers made entirely or mostly of *mitsumata* fiber.

*Mitsumata-shi* has an orange or reddish cast, a gentle luster, and fine grain. It is soft complexioned and rather absorbent, considered the most feminine in feeling of all the various papers. Weaker than *kōzo* and *gampi* due to its shorter fibers, *mitsumata* is, however, more supple and obedient, better lending itself to combination with other paper fibers—in fact, many believe the character of *mitsumata* is greatly enhanced when made in combination with other paper fibers, especially *kōzo* and *gampi*. Therefore, many *mitsumata* papers, notably the modern versions of *torinoko*, *kyokushi*, and *kairyō-shi*, contain other fibers. Old account books of *mitsumata-shi* are frequently collected, crushed along with other fibers, and recycled both by hand and machine. The *Inshū-shi* of Tottori Prefecture and the *hakuai-shi* of Okayama Prefecture are pure *mitsumata* papers, as was the fine quality writing paper (now revived), *Suruga-banshi*.

*Mitsumata-shi* is most often used for stationery, name cards, postcards, *fusuma*, and particularly as decorative paper. Because of their natural luster, *mitsumata* papers are often highly decorated.

*mokuhanga yōshi* 木版画用紙 (local; special; quality)
A *kōzo* paper specifically made for woodblock prints, *mokuhanga yōshi* is a strong and pliant high-class paper synonymous with *hōsho*. It is made in Fukui Prefecture and in the Kyoto area.

*Morishita* 森下 (local; vanishing)
The village of Morishita, believed to be in Gifu Prefecture, long ago gave its name to this paper, which is now synonymous with *kasa-gami*, umbrella paper. A thick, sturdy *kōzo* paper, *Morishita* is mentioned in records dating back to the late fifteenth century.

*Aonami Morishita* was a famous variety from Mino; *Mikawa Morishita* was a fine umbrella paper from the river valley of Mikawa near Obara village in Aichi Prefecture. The modern version has lost the quality of the old paper, using only about 20 percent *kōzo*.

*Nishinouchi (-shi; -gami)* 西の内紙 (local; special)
*Nishinouchi* took its name from the town of Nishinouchi in Ibaragi Prefecture, where this paper is still made. Lord Mitsukuni of Mito was the first to encourage *kōzo* cultivation and papermaking in this area, though there is evidence that *Nishinouchi* may have begun as early as the Kamakura period. Paper merchants brought the product to Edo in the early 1700s to sell. Later it was available in Osaka and other areas. Reputedly *Nishinouchi* ranked with *Mino-gami*, *Suruga-banshi*, and *Sekishū-banshi* in quality.

Originally *Nishinouchi* appears to have been a pure *kōzo* paper, porous, tough, durable, and of a dark cream color. The thicker variety was specially treated and used for raincoats, umbrellas, and *kamiko* (paper clothing). Thinner *Nishinouchi* was used for ballots, *shōji*, deeds, and account books. Down through the centuries the nature of the paper changed. At the end of the Edo period some papermakers began adding rice paste and white clay to whiten the sheets and, when they made handkerchief paper and *shōji-gami* of it, mixed in some *mitsumata* fiber and pulp.

After the turn of the century, manufacturing techniques changed drastically—*nagashi-zuki* replaced the old *tame-zuki* methods to increase production. *Nishinouchi* prospered again for a while during World War II, when it was in demand for paper parachutes and the notorious Japanese paper balloon bombs. After the war the production of *Nishinouchi* suffered a severe decline until 1953 when Ibaragi Prefecture set up its own paper cooperative to aid and teach and encourage its papermakers.

Today papermaking in the town of Nishinouchi is in serious trouble, although paper bearing this name is evidently still produced in some nearby towns and in neighboring Tochigi Prefecture. Its main uses are for *shōji*, woodblock printing, and in books.

*Sekishū-banshi* 石州半紙 (local; special)

*Sekishū-banshi* is believed to be one of the oldest pure *kōzo* papers made in Japan. References to it appear in imperial court records as early as 701. Because of the close proximity of Korea to the western coast province of Iwami, where this paper started, communication between the two countries developed earlier here than with the capital. Some paper historians, particularly Iwami's native sons, hold that papermaking in Iwami predated its introduction to the imperial court. The term *Sekishū* comes from the custom of naming a paper after its place of origin; usually the first character of the province (in this case *Iwa*) is read with its Chinese pronunciation (*seki*), and the word *shū* (area or province) is tacked onto it. *Banshi* is another form of the word *hanshi*, the half-size paper so common to those days.

Lords of both the Tsuwano and Hamada clans promoted the manufacture of *Sekishū-banshi* and eventually monopolized its production. By the eighth century, they were producing more than 180 million sheets a year, making it necessary for every family in the fief, high or low, to grow at least three stalks of *kōzo* in their garden in order to keep the vats fed with raw materials. Obviously *Sekishū-banshi* was in great demand, particularly for account books and important documents. When fires broke out, as they often did, record books were flung into the nearest well. Later they were drawn up and dried in the sun. Because of the paper's superior quality, the dried pages were separate and smooth and, it is said, none the worse for the experience.

During World War II, *Sekishū-banshi* was sent to other parts of the country to be pasted together into balloons—balloons that carried small bombs. These simple weapons were set aloft across the Pacific in hopes of reaching enemy shores. The project was a failure, only a few bombs reaching U.S. land and fewer exploding; but this points out in what high esteem the strength and endurance capabilities of this paper were held.

*Sekishū-banshi* is still made today in the Iwami area, in the town of Misumi, Shimane Prefecture. This fine paper composed of 100 percent *kōzo* fiber, unbleached and hand-beaten, is an Important Intangible Cultural Property, and therefore its production is government subsidized and protected. Today's *Sekishū-banshi* is a deep cream-colored paper with a beautiful complexion, strong and crisp. The thin variety is rather lustrous and fine. Until 1961 *Sekishū-banshi* was also made for the manufacture of *shifu* (woven paper cloth), but evidently no longer. It is still used, however, for account books and documents as well as for *shōji*, stationery, wallpaper, *shikishi* (poetry cards), and for high-quality book papers.

*senka-shi* 泉貨紙 or *senka* (common to vanishing)
Named after a court retainer turned priest-papermaker from Iyo (now Ehime Prefecture) who originated this paper, *senka* appeared sometime in the late 1500s. Its manufacturing process was somewhat unique. The mold was constructed so that its sides could be doubled over, joining two sheets into one thick one. This very thick, stout *kōzo* paper eventually gave its name to similar papers produced in other provinces and was used for books, sutra scrolls, account books, and wrapping. When treated with oil or persimmon tannin, it was often used for umbrellas, for cloaks and raincoats, and as coverings on trunks, notebooks, and numerous other items—a kind of oilcloth substitute.

*Uwa-senka*, once used as a base for sandpaper, later became a popular art printing paper in the West. A *chiri-gami* produced alongside *senka-shi* in Iyo was once made for bags and packaging.

*Senka* is still made in Ehime Prefecture, Shikoku, sometimes under the name *Uwa-chiri*; a type called *Tosa senka-shi* is produced in Kōchi Prefecture, also in Shikoku, and in Kurodani, Kyoto Prefecture.

*Shigarami-gami* 信楽紙 or *Shinano-gami* 信濃紙 (local; vanishing or defunct)
The names for this pure *kōzo* paper came from its places of origin, which include the village of Shigarami near Iiyama along the Shinano River in Nagano Prefecture. Here in various villages dotting the valleys of the northern Japan Alps, papermakers have been conscientiously snow-bleaching and scooping fibers into paper in the same manner for over 350 years, from December through March or April. Often the craftsmen add bits of colored seaweed to the vat.

Their beautiful, white, crisp *shōji* paper was once considered one of the finest made anywhere, this remote area preserving the best of old traditional methods. Although I do not know how papermaking fares there today, it is true that a considerable amount of mechanization and use of chemicals have recently been adopted in this region.

I suspect that this paper was the prototype for several varieties of excellent *shoin-gami* made in northern Nagano Prefecture, including *Uchiyama-shi*, also called *Yama-*

*nouchi-shoin*. *Shigarami-gami* is also quite similar to the *Togakushi-gami* made in the neighboring village of Togakushi.

*Shiroishi-gami* 白石紙 or *Zaō-shi* 蔵王紙 (local; quality)
*Shiroishi-gami* takes its name from the castle town of Shiroishi, Miyagi Prefecture, not far from Sendai. This paper traces its history back to mid-Heian times, making it a very early Japanese paper. A pure *kōzo* fiber paper, reputedly it was a type of *Michinoku-gami*, the prototype for *danshi*.

The lord of Shiroishi, Katakura Kojūrō, did much to encourage papermaking in his fief. Around the year 1600 the Katakura family commissioned expert papermakers from such famous papermaking provinces as Tosa, Echizen, and Uzen to come to the area and to teach advanced techniques—and so *Shiroishi-gami* bears the influence of *hōsho* and *Sugihara* papers. Papermaking flourished here for centuries as the winter occupation of nearly all farmers of the Zao area.

During the mid 1600s, *Shiroishi-gami*, specially made so that all the fibers lined up in one direction, became the material for *shifu*, woven paper cloth. When the fibers were made to cross strongly and squarely, it became the material for paper clothing, *kamiko*, made of crushed and treated paper.

Today only one man has carried on the tradition of making fine, pure *kōzo Shiroishi-gami*, Tadao Endō in the village of Takasu. His paper is used to repair national treasures, for folding screens, *shikishi* (poetry cards), jackets for expensive books, and to a small degree still for the manufacture of *shifu* and *kamiko*. Shiroishi paper was well loved by the modern Japanese artist Gyokudō Kawai, painting instructor to the empress. He called it *Zao-shi*.

*shoin-gami* 書院紙 (common to vanishing)
*Shoin-gami* is the more accurate name for *Mino-gami*. *Shoin* means "study" or "library." Due to its toughness and ability to permit the passage of light and some flow of air into the room, this high-quality *kōzo* paper was sought after for use in studies on the sliding paper partitions, or *shōji*. *Shoin* signifies the highest quality of *shōji-gami*; when it is watermarked with a subtle pattern, it is then referred to as *mon shoin*, or "crest" *shoin*.

During the Edo period, *shoin-gami* utilized the highest quality manufacturing techniques and therefore became famous as the best of all *shōji-gami*. Kimura Seichiku in his *Shinsen shikan* of 1777 wrote: "Among all *shōji-gami*, that of Mino is the best. The purpose of *shoin-gami* is to allow light to penetrate it; therefore the fibers are straightly arranged and the paper carefully made."[2]

*shōji-gami* 障子紙 (common)

*Shōji-gami* is a traditional *kōzo* paper that is used as "panes" when pasted across the latticed sliding doors and windows called *shōji*. *Shōji-gami* is translucent, allowing light into the room but permitting visual privacy. Since the paper "breathes," the room is ventilated even when the *shōji* are shut, and yet the paper retains a room's heat to a surprising degree. Even with the universal use of glass in Japan, many still prefer windows of *shōji*, even at the expense of warmer comfort, for reasons of health as well as aesthetic preference. Many writers speak of the soft, diffused light, which creates an atmosphere of quiet solemnity in a room and harmonizes with traditional Japanese building materials—tatami mats, earth and wattle walls, and unfinished wood. Separate sliding panels of glass and *shōji* are often used together.

*Shōji-gami* is made to fit the size of the latticework *shōji* doors. Papers differ in size, as do the wooden latticed door frames. A sheet 51 × 56 centimeters is standard size; the trimmed sheets are pasted together at the ends into a long roll of about fifty sheets and cut as needed.

In Kurodani I watched a woman making paper to repair *shōji-gami*. A neighbor papermaker impressed on me the difficulty of the work, saying that each sheet of *shōji* paper must be made exactly the same thickness, because when you put it on the same *shōji* frame and the light shines through it, you can see the difference. The woman papermaker laughed and said, "Oh, go on!" Although some margin for human discrepancy is allowed, it is true that *shōji-gami* makers take great pains to make every sheet as uniform as possible in thickness.

*Shuzenji-gami* 修善寺紙 (defunct)

No longer made, *Shuzenji-gami* was at first a wild-fiber *gampi* paper created by a certain Bunzaemon in the late 1500s on a mountain near Shuzen-ji temple on Izu Peninsula, Shizuoka Prefecture. This lovely *gampi* paper with its pinkish complexion and wide chain lines was a favorite of the great warlord Tokugawa Ieyasu, and he patronized its manufacture.

Later when *mitsumata* was discovered as an excellent papermaking fiber, this area so rich in the plant began producing a large quantity of *mitsumata* paper, the skills becoming more widely spread throughout Shizuoka. Eventually *Shuzenji-gami* gained new fame as a *mitsumata* paper beautifully dyed in a range of colors.

An old story frequently told is of the papermaker Shinnōjō, who traveled to another village in order to teach the art of *Shuzenji* papermaking. As a result of his instruction, the house of Aki learned the skills, developed them to a high degree and, producing a portfolio of seven lovely colors, found a noble patron for their paper. On the day, several years later, that Shinnōjō was on his way home at last, a member of the Aki family surprised him on the road and killed him—all so that the secrets of the beautifully dyed papers would never be disclosed.

*Sugihara-shi* 杉原紙 (defunct)
*Sugihara-shi* took its name from its place of origin, but there are differences of opinion as to exactly where that was. Some say it was the Sugihara valley of Satsuma, Kagoshima Prefecture; more likely it was a village in Harima, Hyōgo Prefecture, by the name of Sugihara. Regardless, *Sugihara-shi* was a high-class paper made from *kōzo* and nearly as fine as *danshi*, although not as thick.

*Sugihara-shi* gained public favor and popularity some time around the fourteenth century. Later, several varieties sprang up. The Zen-oriented samurai class particularly cherished it for its simplicity; and during the Kamakura period, when the military came into power, *Sugihara-shi* reigned. Because it made no claim to be ceremonious, it was popular among the common people as well.

Unlike the traditional *Sugihara*, the version produced in this century contained powdered rice for whiteness or was otherwise heavily starched. It is doubtful that it is still made today.

*Suruga-banshi* 駿河半紙 (defunct; limited revival)
*Suruga-banshi* was a *hanshi* paper of pure *mitsumata* fiber once made in the province of Suruga, now Shizuoka Prefecture. Its production centered in particular around the Fujigawa-chō area in a series of rich valleys fed by the Fuji River. *Suruga-banshi* appeared in the late 1700s, soon after the discovery of *mitsumata* as a papermaking fiber.

*Suruga-banshi*, being a *hanshi*, was small. Its color was beigy, its complexion glossy, its nature strong. This writing paper soon became famous and popular throughout the nation during the Edo period, when *hanshi* was the paper favored most for common use.

*tengujō* 典具帖 (common; quality)
The name *tengujō* is probably derived from the name of an important, large shrine (Gujō Hachiman) in the Mino area; since some point in the Muromachi period it has been produced in Mino. Very similar in appearance and character to *urushi-koshi*, although rather thicker and stronger, *tengujō* is made from the finest long *kōzo* fibers, requiring great care in its cleaning and processing. The result is a silky paper of great

fineness, sometimes a mere 0.030 millimeters thick, tough, porous, pliant, and resistant to moisture. The art of making fine *tengujō* is considered one of the highest accomplishments in the whole field of hand papermaking. Much *tengujō* is exported to the West, where it is the favored wrapping material for precious stones, jewelry, and pieces of fine art. It is also used to some extent as a filter paper, for cleaning lenses, and for various medical and dental uses.

By tradition a *kōzo* paper, some of today's *tengujō* is also made of *mitsumata*. Besides being made in the Mino area and in Kurodani, *tengujō* is also one of the predominant papers produced in Kōchi Prefecture, especially in Ino-chō. The *tengujō* made there in the past was excellent; indeed, many who are concerned with the future of paper are worried about this area's paper, for *tengujō* manufacture requires refined and conscientious skills, which are increasingly harder to maintain. Also, the use of *tengujō* is more limited now than ever before, and demands for it have decreased drastically.

*torinoko* 鳥の子 (common)
Traditional *torinoko*, the paper from which legends grew, was originally made of *gampi* fiber only. It was a highly refined paper, so long lasting that it has been called "Japanese vellum." *Torinoko* means "child of the bird," or simply "egg"; in fact, the porcelain-white color and slightly lustrous surface of the old type of *torinoko* does remind one of the shell of an egg.

*Torinoko* is a very old paper, developing from early experimentation with *gampi* fibers. Some *torinoko* made in the Nara period (710–794) is still intact. Literary references to it first appeared about 1330. Its place of origin is unknown, but from about 1480 to the present it has been made primarily in Echizen. At some point in the Muromachi period in Settsu Province (Hyōgo Prefecture), particularly in the town of Najio, a type of *torinoko* was made bearing a significant amount of added clay, giving rise to a whole variety of new clay-containing papers (see *maniai-shi*).

*Torinoko* was quite popular early among the upper classes and, reputedly, expensive. In the middle ages, *torinoko* was sought after for *shikishi* and *tanzaku*, the often gorgeously decorated, standard-sized square and narrow rectangular papers for the writing of poems. This paper was also selected for writing official documents and as the paper for hanging scrolls and *fusuma*. At the time the Versailles Treaty was being put together, officials sought the very finest of papers on which to write this historic document. After much search, two papers vied for selection: English Kent paper and Japanese *torinoko*. *Torinoko* was finally chosen.

The original *torinoko* was probably thin but sturdy, reflecting the finest qualities of pure *gampi* paper. Today it is better known as a strong, thick, stout paper. This more modern version is very white, smooth and somewhat lustrous on one side, close-pored, and available in several thicknesses. It may be made of *gampi* and *mitsumata*, of *mitsumata* and chemical pulp, or even with *kōzo*. The first type is of superior strength and endurance, but the second type, more beige-complexioned, is a good, popular, medium-quality art print paper. *Torinoko-gampi* is primarily used today for stationery and cards, art printing, *fusuma*, and semiofficial documents.

*Toyama-shi* 東山紙; *Toyama* or *Yatsuo mingei-shi* 東山, 八尾民藝紙 (local; common)
These kindred papers of the Yatsuo area in Toyama Prefecture stem from the *hon Takakuma-shi* discussed earlier in the text. They constitute well-made *kōzo* papers whose old techniques of manufacture have amazingly survived the centuries in good health. They are still made today in the same area of their origin. The skills developed here are believed to have been brought all the way from Tosa long ago.

Yatsuo papermakers continue to produce *Takakuma-shi*, as well as a very fine series of dyed papers, which they call *Toyama mingei-shi* or *Yatsuo mingei-shi*, an outgrowth of the former paper. Besides dyeing, artisans also do a considerable amount of lovely stenciled designs on the paper, which finds its way into folk art and gift shops around the country.

*Uchiyama-shi* 内山紙 or *shoin* 書院; *Yamanouchi shoin* 山内書院 (local; special)
In a number of small villages around Iiyama City, Nagano Prefecture, a pure *kōzo shōji-gami* is made that goes by the names of *Uchiyama-shi*, *Uchiyama shoin*, and *Yamanouchi shoin*, all one and the same. This paper was once regarded as being among the finest *shōji-gami* in the country and still enjoys the same fine reputation.

*Uchiyama-shi* is made only in the winter, between December and April, when farming chores are few. Up until some thirty or forty years ago, this paper was a product of locally grown *kōzo* fiber, which was beaten by hand in a mortar and bleached white by laying it upon the snow. Although a few papermakers in the villages of Imai and Tsuboyama still cling to these fine old traditional methods of making *shōji-gami*, others have let themselves become more industrialized, using harsh chemicals, beating the fiber by machine, and using steam-heated dryers—and their paper has suffered. The village of Uchiyama, after which this paper was named, now makes an inferior type of *kōzo-gami* with chemical pulp added.

*Uda-gami* 宇陀紙 (uncommon)
*Uda-gami* is synonymous with *Kuzu-gami* and may be the same paper as that called *misu-gami* as well. Its name comes from a region called Uda, an old center of washi manufacture in Nara Prefecture. Today *Uda-gami* is made both in Ōuda-chō and in the Kuzu-Yoshino area of Nara Prefecture.

During the Edo period, a great deal of *uda* made its way to Edo markets to be sold as umbrella paper, but I am not sure if this is the same sort of paper now made in Nara.

*Uda-gami* today is a *kōzo* paper containing some powdered white clay. It is durable and does not shrink or expand significantly; the addition of the clay makes the paper hang well without curling up at the bottom and so is an excellent base for hanging scrolls as well as screens, *fusuma*, and sometimes *shōji*.

*unryūshi* 雲竜紙 (common)
*Unryūshi* literally translates as "cloud-dragon paper"—for on the surface of this gorgeous washi float hundreds of beautiful, coarse fibers. All *unryūshi* is made of *kōzo*, but local varieties differ in color, coarseness of fiber, and thickness of the sheet.

There are several methods for making *unryūshi*, but two predominate. By one method, the more common one, large *kōzo* fibers that have not been finely beaten are set aside. A strong sheet of paper is scooped, then some of the coarse fibers are added to the solution and the sheet scooped again until a fine layer of the fibers evenly covers the surface. If the coarse fibers are to be dyed, they are beaten first with caustic soda or alum for translucency, rinsed, and then dyed. Alum is added to the vat to make impurities insoluble.

In the second method the coarse fibers are mixed with a bit of vegetable mucilage and placed onto the screen, which has been covered with fine gauze. When a pretty pattern of "clouds" is attained, it is then covered with a thin sheet of paper. The two sheets weld together into one.

*Unryūshi*, unsuitable for writing upon, is a favorite paper for artistic coverings, collage art, wrapping, and for fancy greeting cards. It is made in Kōchi, Tottori, Fukui, and Saitama prefectures, as well as in Eishirō Abe's mill and in the town of Kurodani; in fact, this popular paper is made just about everywhere a large variety of papers is made.

*urushi-koshi* 漆漉
See *Yoshino-gami*.

*usuyō-shi* 薄葉紙

*Usuyō-shi* means "thin paper." Originally the name signified a strong, almost transparent *gampi* paper.

*wara-banshi* 藁半紙 (vanishing)

*Wara-banshi* is *hanshi*-sized (half-sized) paper made with rice straw (*wara*) as its chief ingredient, but also often containing *mitsumata, kōzo*, or other fiber. A very cheap paper to make, *wara-banshi* was and to some extent still is used as calligraphy practice sheets—somewhat comparable to our white newsprint of today.

*Wara-banshi* was once synonymous with *Izumo-gami*, and a very little is still made today in Shimane Prefecture. Also some papermaking towns in Tottori and Yamanashi prefectures produce *wara-banshi*.

*yakutai-shi* 薬袋紙 (vanishing)

Ever since Toyama Prefecture became famous for its manufacture of powdered medicines, traveling salesmen journeyed throughout the country peddling their drugs. During the Edo period they were ubiquitous. The distinctive reddish-brown hard paper used to wrap these medicines on those long journeys became the familiar trademark of the medicine peddler. This was *yakutai-shi*, or "drug bag paper," sometimes called *kusuri-bukuro*. A similar variety made in the town of Takakuma was called *aioichiku-shi*; it was also called *benigara-shi* ("red paper") from the iron oxide used to dye it.

*Yakutai-shi* was made of a mixture of *gampi* and *kōzo*. One side was coated with a protective dye called *go-shiki* ("five colors"), containing persimmon tannin, a waterproofing agent. Reputedly Tosa was the first place to make this paper. A high-quality *yakutai-shi* from Tosa, called *hon-koge*, was used in lower layers of daimyo bedding. As the traveling medicine peddler and his drugs disappeared from the modern scene, so, unfortunately, did the manufacture of his reddish-brown paper.

*Yoshino-gami* 吉野紙; *urushi-koshi* 漆漉 (local; common; quality)

*Yoshino-gami* is not a terribly old paper, but one that made its debut around 1467 in Yamato (Nara Prefecture) in the area of Kuzu after the Ōnin War. Immediately it became a popular pocket paper because it was soft, strong, very white, porous, and, being quite thin and lightweight, flexible. During the Edo period (1615–1868) it was used for napkins, handkerchiefs, and toilet paper.

*Yoshino-gami* is a type of *urushi-koshi*, "lacquer filter paper," for which it is frequently used. Kuzu papermakers still carry on the traditional methods of manufacturing this fine paper, processing the *kōzo* bark carefully. They brush the thinly formed sheets directly onto the drying boards, omitting the pressing, and dry them in the sun. When *urushi-koshi* becomes wet, its evenly interlocked fibers swell up somewhat, closing its pores. This makes it an extremely fine paper.

Two grades of *Yoshino-gami* are made today, one for wrapping cakes and other delicate items and another for filtering lacquer. This paper was used in America for coffee and tea filters until recent times, when it became largely replaced by petroleum papers, although the shortage and expense of petroleum may reverse the demand once more.

### *Watermarked and Water-Patterned Papers*

A WATERMARK is a translucent design or pattern that appears on paper when it is held to the light. The word "watermark" means both the impression appearing in the paper and whatever raised design is fastened onto the screen to make the impression. The concept is simple—when a screen bearing a raised design is scooped into the paper solution, the film of paper fiber on top of the raised section is thinner than the rest of the sheet, and thus the pattern appears faintly on the finished paper.

The history of watermarks in ancient Japan is not well known. Although various extant hemp paper documents of the eighth century bear seal-stamped impressions that look very much like watermarks, in fact they are not true watermarks. It was not until the twelfth century that watermarking is believed to have truly begun in Japan, reputedly in Echizen, but no samples remain today to verify this. Meanwhile watermarks appeared in Europe in the late thirteenth century, probably developing independently. Not, however, until the 1700s did watermarking become a common practice in Japan, mainly as a measure against counterfeiting in paper money issued by the highest offices of government, national and regional, and private banking concerns. As watermarking techniques became increasingly complex and their secrets well guarded, the more difficult they were to duplicate.

Whereas watermarks for handmade papers in the West consist of wire threads soldered onto the metal screens, Japanese watermarks must take into consideration the extreme flexibility of the bamboo and reed-splint screens. Japanese watermarks are made to allow the screen complete freedom of movement and pliability while the paper is being couched. The earliest and simplest watermarks were most likely thread or grass embroidered onto the screen. Today most of the watermark designs one sees are cut from *shibu-gami*, the thick, strong paper treated with persimmon tannin. The delicate design is sewn onto the screen with fine thread, usually silk. Or, the design is sewn or lacquered onto a gauze netting, particularly if the design is large and complex. Then the netting is placed over the screen, or even sewn onto it, before the sheet is scooped.

Another type of watermarking involving a truly remarkable technique is *kurozukashi*, or *kurosuki-ireshi*. Its skills are secret, protected by law for the exclusive use of the government's Printing Bureau. *Kurozukashi* is a kind of watermarking in reverse, where the pattern is cut out of the screen rather than applied on top of it. A *mitsumata*-based *kyokushi* is always used, which must be made by the *tame-zuki* method. Developed in the latter part of the nineteenth century by the Ministry of Finance's Printing Bureau, *kurozukashi* is a complex watermarking technique employed today only by the government for bank notes and official documents.

In contrast to *kurozukashi*, which is a fairly recent, if secret, development, regular types of watermarking were well known and commonly practiced in old Japan. Soon after watermarking's inception, papermakers everywhere were creating all manner of beautiful, decorative papers utilizing allied techniques. By adding color and using controlled channels of forced water, they created many new papers with totally new effects. Among these are *mizutama-shi*, *arare-gami*, *unjō-shi*, *suiryū-shi*, and *rakusui-shi*.

*Mizutama-shi* means "drops of water paper," and that is literally how the design is achieved. Originating in the eighteenth century, *mizutama-shi* was made in various provinces, notably Echizen, Harima, and Higo (Fukui, Hyōgo, and Kumamoto prefectures). At first the base was a clay-containing *gampi* paper used for decorative book covers, but later it came to be made of *mitsumata* or *kōzo*. As the story goes, the Iwano family in Echizen made *mizutama-shi* generations ago in this manner: a sheet of paper was formed on the mold, then a leaf dipped into water was flicked over it. Where the drops fell, shallow holes were made in the sheet, just deep enough to create a pattern but not so deep as to penetrate through it. Maintaining control over an even pattern was quite difficult.

Various methods for making *mizutama-shi* are used today, all more efficient. In one method a piece of metal sheeting cut into an even pattern of holes is placed over a sheet of wet paper. A hose with a sprinkler attached is moved across the sheet, releasing a gentle spray of water. As the water falls through the "stencil" holes in the sheeting, dots are created in the paper. Either this patterned *mizutama* is used as is, or else the patterned sheet is dyed and placed on top of a white sheet. Since the sheets adhere when dry, the result is a colored sheet bearing soft white dots. Because the sheets pelted with water are fairly fragile, the manufacture of *mizutama* is a tricky procedure.

*Arare-gami*, meaning "hailstone paper," is essentially the same type of paper as *mizutama-shi*, and other variations abound. In *unjō-shi*, two thin sheets of *kōzo* paper, one dyed and one white, are made with extra mucilage for transparency and fineness, then patterned with water and joined as in *mizutama-shi*. In *suiryū-shi*, a pleasing pattern resembling exaggerated chain lines of a screen are artificially created by running channels of water across a colored sheet, which is then joined with a plain white sheet. In *rakusui-shi*, just one thin sheet of *kōzo* or *mitsumata* paper is made. On top of this is placed a patterned "stencil" consisting of either wire mesh or gauze; cutouts of treated paper, tin, or other metal; a piece of lace or openwork cloth. A strong spray of water is forced through the stencil and all the way through the sheet, opening up a delicate pattern of holes to create this "lace" paper.

### *Dyed Papers*

*zenshi-zome*—dyeing the whole sheet 全紙染

THE Koreans are believed to have been the first people to have dyed their papers. Using plant dyes, they colored the paper fibers thoroughly before the solution was scooped into paper. Not much later, the Japanese began devising ways to color their papers, a natural preoccupation for these people, who have always sustained a tremendous zeal for dyeing and have raised it to a great art—in fact, bright red, blue, and yellow hemp paper of unusual clarity of color manufactured in the middle of the eighth century is part of the collection of the Shōsō-in, the Imperial Repository in Nara.

The same vegetable dyes anciently used to dye clothing and textiles were used to dye paper. Roots, bark, leaves, berries, flowers, fruit skins, nut shells, and galls most frequently made the bases of dye solutions. Some of the most popular dye plants were *ai* (indigo); *kihada* (Chinese cork; *Phellodendron amurense*, Rupr.); *yamamomo* (*Myrica rubra*); *hamanasu* (sweet briar); *kuchinashi* (Cape jasmine); *murasaki* (*Lithospermum officinale*); horse chestnut leaves; *suo* (sappanwood); and walnut skins, to name a few. The powdered mineral *benigara* (red iron oxide) was also used extensively for dyes. Papermakers kept detailed notebooks in which they included samples of colored papers, drawings of plants, dye recipes, and records of which mordants produced different color tones from the same basic dyes.

The papers were dyed by one of three possible methods: *suki-zome*, *tsuke-zome*, or *hiki-zome*. In *suki-zome* the raw fibers are dyed before they are made into paper. The fiber is beaten, squeezed of excess moisture, and placed in the dye solution for one day. Then it is removed from the dye and placed in the mordant for at least half a day more. Sometimes the dyeing and mordanting process is repeated. Then the thoroughly dye-saturated fibers are placed into the vat, mixed with water and mucilage, and scooped into paper.

In the *tsuke-zome* method the formed paper sheets are immersed in a liquid dye. The dye solution is somewhat diluted, and the paper drawn through the liquid, then through the mordant, and dried. This procedure is carried out three times. After the final drying, the sheets are rinsed in clear running water for about an hour. Repeated dyeing of the sheets in a weaker solution gives a better and more durable color than one dyeing in a strong solution. A third method of dyeing paper is called *hake-zome* or *hiki-zome*: a soft, wide brush is dipped into the color and spread over one or both sides of the paper until it is thoroughly saturated.

*kon-shi*—indigo paper 紺紙

The old and distinguished art of making *kon-shi* has all but died out. This strong and durable dark blue paper dyed with indigo was produced just before and during the Heian period for copying sutras. The Buddhist scriptures were gracefully inscribed, and often paintings were done, on the indigo paper with gold or silver ink, producing works of striking beauty and, for some, a sense of profound spirituality and mystery.

*Kon-shi* utilized a rather good quality *gampi* paper, which has some natural water-resistant qualities. The paper was soaked in an indigo solution mordanted with lye, then removed and dried, and the whole process repeated seven to eight times before it was considered sufficiently dark. Then the sheets were soaked in running water for one day and washed to remove the last traces of insoluble indigo and lye, which would stain or muddy the color. The dried *kon-shi* was then beaten somewhat with a mallet to give it a fine surface and sheen and glazed with an alum and glue sizing. Making *kon-shi* required not only time but skill in the proper removal of dye impurities.

Some *kon-shi* is still produced in the Kyoto area, but its manufacturing techniques have been short-cut; for example, the dyes are mixed with the paper fibers before scooping. The result is a *kon-shi* lacking the depth of indigo color of the old *kon-shi*. A good indigo paper is made in the Tokushima area of Shikoku.

*kata-zome*—stencil dyeing 型染

*Kata-zome* is a technique applied only quite recently to the dyeing of paper, at least to the extent seen today, although it has been used since at least the eighth century in dyeing textiles, and stencils also had some application to paper in the Edo period, including a type of *ukiyo-e* print known as *kappa-zuri*. Stencils have been used from about 650 A.D. in China. At the Tun-huang cave temples in northwestern China, thousands of stenciled images of Buddha lined the walls and ceilings of a sealed monastic library cave. The first Chinese paper stencils were simple designs made by pricking paper with pins. This design concept is still very much alive in Japan, but the tiny, perfectly circular holes today are cut with sophisticated bladed tools.

Japanese polychrome *kata-zome* reached its peak of development in *bingata*, a method of stencil and paste-resist textile dyeing that originated in Okinawa. *Bingata* traces its roots back about three hundred years, to Japanese *Yūzen* stencil dyeing and to Chinese techniques, but known *bingata* examples are not older than two hundred years. True *bingata* from Okinawa is characterized by its bright, yet subtle color schemes, particularly pinks and reds, blues, and yellows, and certain recurring design motifs.

The process of *kata-zome* on paper is as follows: specially treated stencil paper (see Treated Papers) is cut into a "positive" pattern with a stencil knife. The stencil, which if delicate or complex has been strengthened with silk gauze, is placed on top of the paper. A thick, creamy paste made of rice bran, rice flour, salt, and slaked lime is spread over it with a spatula. This starch resist fills all the negative areas. Then the stencil is carefully removed. Many runs can be made, and a continuous pattern is possible. When the resist has dried on the paper, the whole sheet is coated with a soy bean milk, which is a protein and acts as a mordant. Then the paper is allowed to dry. Colors—dyes or pure pigment solutions—are mixed and gently brushed into the areas not covered by the resist. Generally, several colors are applied at once in various open areas. When the colors are completely dry, the paper is floated in a sink of water until the resist begins to loosen and melt away. The loosening may be promoted by nudging with a feather or by repeated rinsing. Miraculously, the colors remain on the paper in a design.

The modern application of *kata-zome* methods to paper was an inspiration born of necessity. During World War II, cloth was extremely hard to come by in Japan, especially for purely artistic purposes. Artisans Keisuke Serizawa, Seikichirō Gotō, and others under the encouragement of Sōetsu Yanagi—who had just returned from Okinawa enthralled with the *bingata* designs there—began experimenting with *kata-zome* on washi with fantastic effects. The result is the many beautiful and delightful prints and patterned papers that have emerged since the war, most showing the folk art influence. Top *kata-zome* artists working with paper today include Serizawa, Gotō, Sadao Watanabe, and Yoshitoshi Mori. Serizawa was designated a "Living National Treasure" in 1956 for his *bingata* techniques. Much of the *kata-zome* art found in folkcraft shops all over Japan originates in Serizawa's Tokyo workshop, from the papermaking town of Yatsuo in Toyama Prefecture, and from Kyoto area workshops.

*shibori-zome*—tie dyeing 絞染

Japanese tie dyeing, or *shibori-zome*, is an art even older than starch-resist dyeing methods. Although the tie dyeing of washi does not have the gentle subtleties of the traditional *shibori-zome* on silk and cotton, it is colorful, gay, and popular. Because of the rough treatment the paper must withstand, only *momigami* is used, a heavy, strong washi that has been treated with *konnyaku* (devil's tongue root) starch and wrinkled over and over again by hand (see p. 205).

To begin *shibori-zome*, a female worker lightly marks dots in pencil on the sheet, indicating areas to be tied off. Then she dampens the *momigami* with a towel moistened in either water or *konnyaku* solution, just enough to lend the paper pliability and "give." Taking a long pointed tool, such as a child's chopstick, she places the tip at the dot and draws the folds of paper down over it. With thread or string she begins tying around the paper point, drawing away the stick after about two complete twists, and continues wrapping the folds firmly for the desired number of inches. She may then gather a small group of points together, and wrap them as one. The ends of string are tied in slip knots, which come apart promptly when tugged. The sheets are immersed in a dye vat, agitated for thirty minutes, rinsed, then hung on a line to dry. When no longer dripping wet, the strings are pulled off.

*Shibori-zome* done on washi is a specialty of Kurodani, Kyoto Prefecture, and it is done in a few other papermaking areas as well. These tough, decorative papers are made into paper rugs, obis, cushion covers, book covers, wallets and purses, and many other useful objects.

*Itajime-zome*—fold and dye 板〆染

*Itajime-zome* is a technique used in a similar manner to dye both paper and cloth, although the patterns produced in paper are generally more sharply defined. The process is simple: a thin but strong paper is selected and folded into any one of a number of patterns, the last folds pleated into an accordion shape. The folded paper is dipped into water until saturated, then removed and squeezed gently to remove all extraneous moisture. Sometimes the folded sheets are bound with rubber bands, sometimes not. The damp paper is dipped by its corners or sides into dyes, often one color on top of another. Then the sheet is opened carefully and dried. In traditional *itajime-zome*, two wooden blocks carved into symmetrical patterns bearing slits or holes are placed at either end of the folded cloth and clamped together. Either the blocks are dipped into the dye vat, or the dyes are poured through the openings, or both.

The women of Kurodani produce lovely fold-and-dye papers today. They are also made at Eishirō Abe's mill and in other papermaking towns in the north.

*sumi-nagashi*—marbling 墨流

*Sumi-nagashi* means "ink flow"; it is the art of creating one-of-a-kind marbleized designs on paper by touching the surface of a basin of water with a brush dipped in *sumi* or colored inks; the paper is placed over the design and pulled away. The West generally attributes the Persians with the invention of marbling, pinpointing its beginnings around the fifteenth century; but possibly these gorgeous patterns, which are produced in both East and West by similar methods, originated with the Japanese much earlier.

Although scholars are uncertain when and how *sumi-nagashi* first came about, they know it was already a highly developed art in Japan by the twelfth century. Specimens of *sumi-nagashi* that date back to late Heian times are preserved today in various collections and museums in volumes of the *Sanjūrokunin-shū*, the poetry of the Thirty-six Poetical Geniuses of Heian. Especially popular among the aristocracy during the Heian period, such patterns then did not fill an entire sheet, but appeared in miniature only in the right-hand corner of a page, the rest left free for poetry. The Hiroba family of Echizen (Fukui) have practiced and passed down *sumi-nagashi* skills for centuries; indeed, Hiroba family legend relates that the art was a divine gift bestowed upon them by the god Kasuga in 1151. Toward the end of the seventeenth century, the Uchida family also became famous as makers of fine *sumi-nagashi*. During the Edo period, gorgeous *sumi-nagashi* appeared, which had patterns on both sides and was colored with gold dust. The art flourished until World War II, when it suffered many of the same setbacks of other arts and crafts in Japan. In the more prosperous years following the war, *suminagashi* revived and continues to be made in Fukui Prefecture by the Hirobas.

The *sumi-nagashi* method is as follows: generally *hōsho* or *torinoko* from Fukui is used. The craftsman holds four brushes in his hands, the left holding three dipped in color, the right a brush dipped in water and perhaps slipped through his hair to obtain a little oil. The craftsman stoops over a wide enamel basin filled with water. Poising himself, he suddenly pokes one point of color into the center of the water's still surface. Without hesitation he cuts into the center of the color point with the uninked brush—immediately the color breaks into an outward flowing ring, like ripples in a pond. Poking again with color, then breaking it into a ring, over and over, alternating colors, he quickly forms one hundred or more erratic, concentric rings that have spread over the sheet to the basin's edge. Then, leaning to one side, he gently blows across the surface

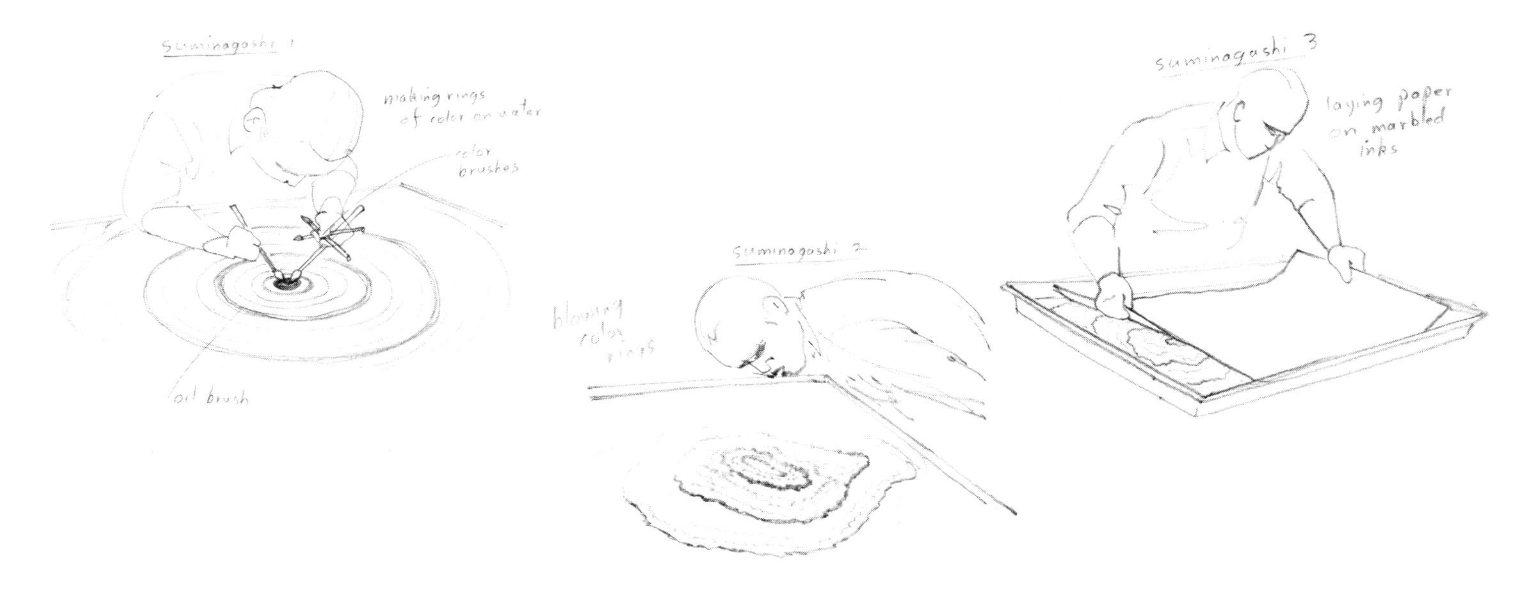


of water at an angle, perhaps also fanning it into shapes or moving the design about with a stick. All his movements are performed with extreme care so that the rings will not break into each other or the colors muddy the surface. When he has achieved a pleasing whorled pattern, the artisan takes up a piece of paper and, holding the short ends up, carefully lays it, center first, upon the water's marbled surface. As the water saturates the paper, the design penetrates through to the back of the sheet. At this point he takes in hand a flat stick, slides it under a long edge of the sheet, and carefully lifts it away. The colors used in today's *sumi-nagashi* are generally *sumi* black along with red and indigo dyes. Primarily, sheets of *sumi-nagashi* are used for writing poetry, but recently it has also come to be used for endpapers, book covers, and other decorative uses.

### *Sukiire-gami—Imbedded Papers* 漉入紙

IN ANCIENT days, it is said, papermakers sometimes threw bits of their own hair onto the paper screen as they scooped their sheets, to identify them as their own. If so, these were probably the first intentionally imbedded papers. By the middle of the Heian period, papermaking had developed to a considerable degree of technical excellence; not only were papers elaborately colored and decorated, but the designs became progressively more sophisticated, as the tastes of the imperial court demanded. Elegantly decorated sheets and cards for the writing of poetry were among objects most cherished by this highly refined society in its passion for the written word. Sheets appeared imbedded with flakes of metal foil and threads and strewn with gold and silver leaf and dust for their shimmer—or mica, sand, alum, and talc powders for their soft luster.

When residing in Yoshino during the 1300s, Emperor Godaigo ordered the manufacture of poetry cards made of *otaka danshi* imbedded with ivy leaves. This was probably the first time organic matter was purposely inserted among paper fibers for its decorative effect. After that, poetry cards bearing all manner of plant life appeared, suitable for writing poetry appropriate to the seasons. In the Edo period, invitation cards for such events as Nō drama parties given by the court contained fanciful bits of nature.

Fine grasses; colored pieces of seaweed; cherry, plum and apricot blossoms; pine needles; leaves and leaf skeletons; threads; bits of bark *chiri*—all came to be imbedded in papers for their gorgeous effect and were widely appreciated at court and in literary circles into the early 1800s. Insects and butterflies inserted in paper was the invention of the Nawa Museum of Entomology in Gifu in recent years. Nagano and Gifu prefectures today produce most of the *sukiire-gami* now available.

*Treated Papers*

UNDER this group fall four major types. The first comprises those papers treated with persimmon tannin (*shibu*), in particular *kata-gami* for use in stencil arts and *fuda-gami* for dyeing textiles. Papers treated with a mucilage made from devil's tongue root (*konnyaku*) come next; these include wrinkled *momigami* and the rich variety of paper clothing known as *kamiko*. Those papers treated with oil form another important group deserving at least brief mention, comprising papers used for umbrellas or as a kind of oilcloth. "Leather" papers and the *danshi* paper are categories unique and extraordinary in themselves.

When the Japanese began treating their paper with protective agents, they imparted to it qualities that made paper much more than just paper. By endowing it with increased strength and durability, by providing water and insect resistance, and giving paper additional flexibility or rigidity, they changed its rather simple and limited nature, making paper one of the most versatile of all modest materials. From ancient times down to the present, treated papers became protective coverings against rain and snow, from clothing and rainwear to huge pieced-together sheets screening crops, store front goods, or anything that had to be transported openly in wet weather. Treated papers were essential for the conduct of business as bags, used over and over through the years and covered from top to bottom with tiny paper patches. Beautiful and highly durable wallets, boxes, dishes, cannisters, sword sheaths and even bottles were fashioned of strong paper and coated with lacquer. It is astounding to consider these objects of such beauty and imagination—often born of necessity and economy—that have been fashioned from paper on these islands from ancient days down to the present.

*Shibu-Treated Papers*

Called *shibu-gami*, this *shibu*-treated paper is of two major types, *kata-gami* (stencil paper) and *fuda-gami* ("label" paper). Although some *kamiko* is also treated with *shibu*, largely it is treated with *konnyaku*, and so falls under that heading.

*Kata-gami* is believed to have originated in the southwest area of Korea known ages ago as the kingdom of Koguryŏ (A.D. 37–668). The paper's date of appearance in Japan is uncertain, but it is undoubtedly quite early, since stencil dyeing has been practiced in Kyoto alone for over one thousand years. The center of *kata-gami* manufacture historically and today is Shiroko-chō in the area of Ise, Mie Prefecture. Town records speak of four *kata-gami* makers producing these papers there at the turn of the ninth century. According to Shiroko legend, an old man who lived in a temple there in ancient times was cleaning the temple grounds one day when he noticed a leaf on the grass, eaten away to a skeleton by bugs. Surprised by the power of nature, he brought the leaf home and contemplated it as it lay on his table; regarding its fragile bony pattern, he conceived the idea of using it as a stencil for dyeing. He placed it on a sheet of washi and cut the paper through the leaf openings, so originating *Shiroko kata-gami.*

In the Kamakura period, warriors began favoring kimonos printed in stencil designs, especially the high-ranking samurai. But during the three hundred years or so of war following this period, the art declined, since fewer and fewer skilled artisans continued to keep it alive. In Edo times, stencil dyeing again revived, and *kata-gami* was in great demand. That produced in the Shiroko (Ise) area today has been named an Important Cultural Property, and there are three "Living National Treasures" in Ise as of this writing involved in aspects of making *kata-gami.*

The general method of making this stencil paper is as follows, although techniques differ from place to place. Thick *kōzo* or *mitsumata* paper from Hiroshima Prefecture or the *Ise kata-gami* of the Mino area of Gifu is most commonly used as a base. The white sheets are cut into standard sizes, and three sheets, one on top of another, are fused together by brushing them with persimmon tannin (*nama-shibu* or *kaki-shibu*), the processed juice of unripe persimmons. The sheets are then separated and dried on pine drying boards in the sun. After that, the coated papers are smoked in a smokehouse for about ten days, a wood fire kept smoking beneath them throughout the day. This gives the paper its characteristic smoky fragrance. After coating the sheets once more with *shibu* and drying them, the *kata-gami* is aged from two to three years. Sharp lines and shapes can be cut into this thick, nut-brown paper with stencil knives; its hairs do not feather out as much as other washi, yet it is exceptionally sturdy, wet or dry, and not too slippery, making it an ideal stencil paper particularly well suited to stencil-resist dyeing methods.

*Fudagami*, literally "label paper," is a rather thick paper, usually of *kōzo*, which has been treated with *shibu*. Its main use is in labeling and marking textiles during dyeing.

*Fuda-gami* has strength and good water-resistant properties, and that is why one of its thicker varieties was used in the manufacture of umbrellas; also it was once essential to the silkworm industry. A certain smaller variety goes by the name of *shibu fuda-ban*.

*Fudagami* is produced today in Kurodani, where the paper is made and treated, and in Kyoto, where various companies bring in good-quality papers and treat them with *shibu* to make *fudagami*.

### *Konnyaku-Treated Papers—Momigami* 揉紙

*Konnyaku* is a tuberous root of the devil's tongue plant of the *Arum* genus, the starch of which is used to produce a jellylike substance common in the Japanese diet but also useful as a starchy mucilage. *Konnyaku* mucilage has been used for centuries to coat papers, particularly in the manufacture of *kamiko* paper clothing (see p. 53). When a thick, good quality *kōzo* paper is treated and repeatedly crumpled by the hands it is called *momigami*, or "kneaded paper." Such paper is often given greater strength by treating it with *konnyaku*. When made into clothing, it is referred to as *kamiko*.

The application of *konnyaku* makes paper strong and flexible enough to withstand the rubbing and wrinkling process; both treatments, in turn, render *momigami* much stronger, softer, and more flexible than untreated paper. Because the mucilage coats the paper's pores, the sheet becomes not only wind- and water-resistant, but the paper's natural heat-retention qualities are enhanced; and yet the paper still breathes. Sheets of *momigami* can be dyed or left plain; then they are glued end to end to form a roll or bolt and sewn into clothing, particularly coats. As lining for clothing, *momigami* often outlasts the textiles that it lines.

The method for making *momigami* is this: the worker prepares the *konnyaku* by boiling the vegetable or its concentrate in water to obtain a gray, viscous solution. He then sponges the *konnyaku* solution over both sides of a sheet of paper and, folding the four corners into the center, crumples it gently into a loose ball. Placing the balls onto a sheet of plastic until a substantial number have accumulated, the worker takes them up one by one, packs each into a ball, turning it round and round, squeezing and wrinkling it carefully but firmly. He then unfolds the sheet, rewraps it, and repeats the wrinkling and crumpling. After three to four minutes of this, the worker opens up the sheet and, grasping the near corners, rubs them together between his palms, paper rubbing against paper. He rubs down the entire sheet, then lays it on a flat surface, stretching it slightly by applying pressure with his hands in an outward direction toward the sides and corners. All the crumpling, rubbing, and stretching may then again be repeated.

*Nishinouchi*, *senka*, and *kyōsei-shi* ("strongly made paper") are names of thick, high-quality *kōzo* papers traditionally used for the *momigami* of *kamiko* paper clothing—all specially scooped and stroked on the mold so that the fibers cross firmly at right angles for utmost strength. Its unusual wrinkled texture makes it an interesting paper for certain kinds of art printing. *Momigami* not treated with *konnyaku* is used to polish lacquer ware. *Momigami* is made today in Shiroishi, Miyagi Prefecture; in Kurodani, Kyoto Prefecture; and in a number of other northern papermaking towns.

*Oiled Papers*

A short note only on this very useful, once indispensable paper, which has all but disappeared from Japan. Good-quality paper, usually of *kōzo*, combined with good drying oil properly applied was the secret of fine *abura-gami* (literally, "oil paper"), and rather little is produced today. *Abura-gami* was to old Japan what plastic sheeting and waterproof synthetics are to the world today. Used over and over again, and patched wherever torn, oiled paper served as emergency raincoats, as bags for storing tea and grain, as sheeting to protect crops from frost, for covering goods displayed outside a shop, and for paper lanterns that resisted wind and rain when carried by a cross-town or cross-country traveler on a stormy night. Its use now is mainly for umbrellas: good-quality washi is pasted to the bamboo ribs, treated with *shibu*, then coated with drying oil for resistance and to lend some pliancy to the paper. *Abura-gami* is made mostly around the Kyoto area and in country villages by farmers for their own use.

*"Leather" Papers—Kinkarakawa* 金唐革; *Gikaku-shi* 擬革紙

Originally these leatherlike papers arose during the Edo period when Buddhism discouraged the slaughter of animals. Thick, wet washi cardboard was pressed into coarsely woven textiles or mats to emboss the sheets; while still damp the sheets were shaped around molds or formed freely into container forms, which were oiled or lacquered. During the nineteenth century, the embossed patterns became elaborate and unique, and the painting with lacquer became a high art in the manufacture of *kinkarakawa-shi* (see p. 142).

*Danshi* 檀紙

*Danshi*, the elegant, thick, snow-white furrowed paper so beloved by the nobility evolved from *Michinoku-gami* (see p. 183). For an extensive time throughout Japanese history the names of the two papers were synonymous. Originally *danshi* did not have the deep furrows it possesses today, although it may have had at least a slight natural wrinkle to it. It was most probably made of *kōzo*. Some records indicate that it was not wrinkled until about 1700, although other historians say it may have been furrowed as early as the year 1000. At any rate, the literature of the time reveals that this paper, still referred to as *Michinoku-gami*, became thicker as years passed and took on a yellowish cast; and the unique method of manufacturing it was highly secret. It was used largely for ceremonial purposes, such as writing, wrapping gifts, and as a gift itself.

This paper became so popular that in the centuries following its origin imitations naturally appeared. The collective name for all of these similar papers was *danshi*. *Danshi* means "sandalwood paper"—probably so called because of its unique ability to retain fragrances; in fact it was often perfumed with incense and used as a handkerchief or towel. In the 1300s the name *hiki-awase* found its way into usage, referring to a highly superior and tough kind of *danshi* used at that time to line armor. By the mid 1600s this name, too, began to disappear.

Today's *danshi* is inferior to that of old but is still considered, if unofficially, the most ceremonial of Japanese papers. The Yanai family of Takahara, Okayama Prefecture, who once made paper for the powerful Tokugawa shogunate, now make *danshi*, and it has been appointed a Cultural Property of the city. A kind made partly of pulp is also produced today in Ehime Prefecture. Tadao Endō of Shiroishi, Miyagi Prefecture, produces a type of paper resembling *danshi*, which he calls *chijimi hōsho*.

Sanuki and Bitchū provinces (Kanagawa and Okayama prefectures) became the chief producers of the new *danshi*. It was still made chiefly of *kōzo*, but after the Kamakura period (1185–1333) its character underwent some changes, notably becoming whiter and glossier. *Danshi* from the Bitchū area was considered the finest and noblest, being in great demand in Kyoto and Edo (Tokyo). From the end of the 1500s until the nineteenth century, this area's *danshi* was authorized by the imperial court and the Tokugawa shogunate as the government's official paper, used inside shrines and by the ruling families.

Often called "creped" paper, *danshi* has also been known historically as *mayu(mi)-gami*, *shohi-gami*, and *mutsu-gami*. Today three types of *danshi* are made: *ōtaka*, *chūtaka*, and *kotaka*, describing varieties ranging from the thickest, most wrinkled kind to the thinnest, least creped *danshi*. The manufacturing method is as follows: four sheets of *kōzo* paper containing lots of mucilage are made and stacked upon a board, each sheet being moistened with water and pressed with a spatula. When these four sheets are very slowly and carefully pulled from the board before they have dried, deep furrows form

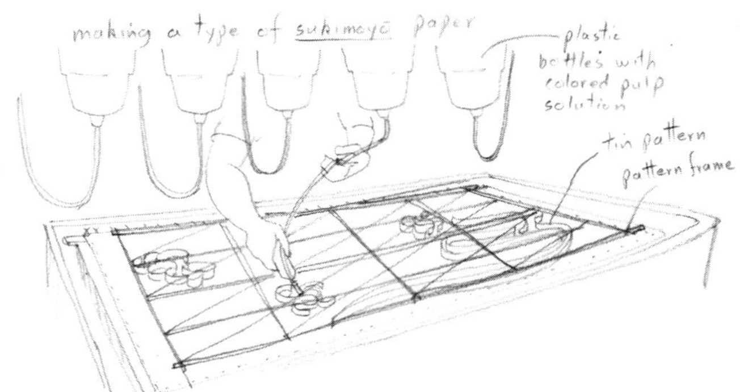


on their surface. The more sheets used, the deeper the ridges. The top sheet only is dried, the rest used again to create texture in another sheet. *Ōtaka*, *chūtaka*, and *kotaka* originally referred to the size of the sheets but later came to designate the depth of the paper's furrows.

### *Sukimoyō-gami—Elegantly Designed Papers* 漉模様紙

THESE designed papers constitute the most ornate papers imaginable. Most originated in the late Nara and Heian periods, when the court and aristocracy demanded gorgeous and imaginatively opulent papers upon which to write poetry and to copy Buddhist sutras. Papermaking was a newly evolved craft, and there was a joy in experimenting with color and texture. Designers of imaginative papers took delight in helping influence the tastes of court poets.

Elegantly designed papers not only continue to be produced today, but every year more are invented. One design by Eishirō Abe has even been patented. The group of papers referred to as *sukimoyō-gami* include *tobikumo*, *ungei-shi*, *uchigumo*(*ri*), *bokashi*, *hikikake-gami*, *sukie*; and *kumo-gami* and *ramon* not described here. Also, *unryū-shi*, *mizutama-shi*, and *rakusui-shi* discussed earlier are sometimes included in this category. These ornately designed papers were used for long, narrow *tanzaku* and square *shikishi* poetry cards early in their history and later were used for *fusuma* paper, for the covers of kimono boxes, and for some hand scroll and hanging scroll mountings.

*Tobikumo* means "flying clouds." Originally this paper had coin-sized blue dots of fiber scattered over it, but the *tobikumo* that has come down to us had clouds of purple and indigo fiber floating across the sheet. The method is basically this: a sheet of plain white paper is formed. In a pan placed near the vat the papermaker mixes colored fiber with water and a little bit of *tororo-aoi* mucilage—but not so much as to ruin the color. Taking the pan in his hands, he tosses the mixture over the formed white sheet in a pleasing, free-form pattern. Another very thin sheet of white paper may or may not then be formed over the whole work to protect the colored fibers from peeling off.

*Ungei-shi* is made by a method very similar to *tobikumo*. Plain, long, thick *kōzo* fibers are mixed with a bit of alum and mucilage, then poured into a swirled pattern over the base sheet as it sits in the mold. Tadao Endō of Shiroishi makes a beautiful variety called *haku unshi*, using base papers colored with vegetable dyes.

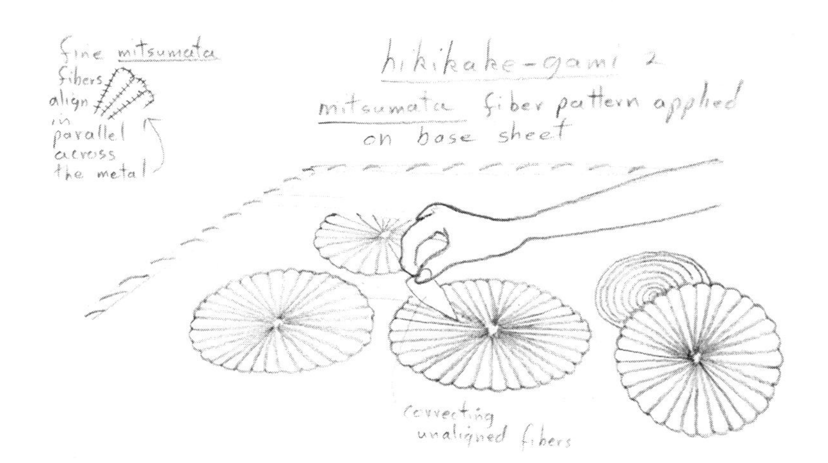


*Uchigumo(ri)* is a very old paper, originating sometime in the eleventh century, when it was used for poetry cards. A base paper, traditionally *gampi*, is formed and set aside. Some indigo-dyed paper is then beaten with water and perhaps some mucilage in another vat. Then the white *gampi* paper in its mold is dipped into the vat of color along the bottom edge, so that an irregular pattern of color flows over the sides. Again the mold is dipped, and color overlays color. The same movements are performed with the opposite edge of the sheet. Purple-dyed paper is often used along with blue. The *uchigumo* of Heizaburō Iwano of Ōtaki, Fukui Prefecture, is an Intangible Cultural Property.

Variations on these "cloud paper" patterns include placing various dyed fibers mixed with mucilage in containers such as metal boxes or flexible plastic bottles. The colored fiber solution is poured mechanically or by hand across the length of the base sheet, producing a thin wash of colored fiber over the paper. Another method utilizes a frame equipped with small metal chambers each holding a solution of dyed fiber; when the top of the frame is flipped over the base sheet, color flows over the base in a streaming pattern. Then the screen is shaken some. Or, alternatively, a patterned sheet made by this method is overlaid upon a white one. In other variations, partitioned metal forms of elaborate design impart pattern and a variety of color when they are placed over a base sheet of paper and dyed fibers poured into the partitions. In fact the variations possible by this method are unlimited.

Another process particularly prevalent in Fukui Prefecture paper mills is called *hikikake-gami*. A tin mold formed into a pattern, resembling soldered-together cookie cutters, is lowered into a solution of *mitsumata* fibers, then lightly impressed upon a silk *sha*, or gauze, floating on top of a vat of water. Magically, the glossy, patterned *mitsumata* fibers transfer onto the *sha*, which is then inverted onto a colored base sheet.

By another very delicately executed method, a skilled papermaker can make what is called *moyō-gami* ("patterned paper") or *kōshi*. Generally *mitsumata* fibers are used because they are relatively short and fine; powdered mica or alum may be added to them for soft luster. A delicately patterned sheet is formed of *mitsumata* fibers, usually dyed, by placing a stencil over a *sha* and scooping. The fibers settling on either the stencil or the *sha* are carefully transferred to a base sheet, probably of *torinoko*.

A type of patterned paper that resembles cloth or canvas in texture, particularly the *nunome-gami* (or *-shi*) produced in Toyama Prefecture, is a thick *kōzo* paper pressed onto a highly textured piece of cloth just after it is scooped. A paper similar to this was made in the southern islands of Japan before the war and, after being treated with a special coating, substituted as canvas for art work.

*Bokashi* paper was possibly one of the earliest designed papers. A pattern of gradated or shaded color was obtained by combining various methods of one-color dyeing. The sheets were dip-dyed along their long edges; then more color, perhaps a contrasting one, was brushed over this, creating flowing edges of shaded hues.

*Momigara-gami*, meaning "crumpled-pattern paper" is a specialty of Shiga Prefecture, an area that produces many clay-containing papers. *Momigara-gami* is loaded with so much clay that, when a colored sheet is wrinkled in the hand, a design of batiklike crackles results. Its main use is as endpapers for traditional Japanese-style books.

*Sukie* includes all those papers that are literally "painted" with colored fiber. Papermakers turned "artists" in Obara-mura, Aichi Prefecture, and in Ōtaki, Fukui Prefecture, among other places, have devised ways of creating novel and sometimes lovely pictures, realistic and abstract, by applying colored pulp and fiber to base sheets of paper. In my own prejudiced opinion, these works have little connection with the craft of making fine papers, but they are an interesting offspring that deserve note here.

### *Nonpaper Papers*

Two noteworthy types of what I would have to call nonpaper papers are made in Japan. One is the veneer paper sometimes called *kiri*, *sugi*, or *mokume* ("wood grain pattern").

*Kiri* is paulownia and *sugi* is cryptomeria, the Japanese cedar. These are some of the trees from which extremely thin slices of the woody core are cut by hand. The ends of the veneer are butt-joined and bonded together on top of a base paper sheet. These veneer papers, though not true paper, are nevertheless lovely and fascinating. They are often used for name cards and art printing in Japan. Many of these veneer papers cut from various woods are available in the West.

Another nonpaper paper, although not available commercially, is the tapa cloth made by Tadao Endō of Shiroishi. Endō became fascinated with the traditional tapa and kapa of Polynesia. Discovering that the mulberry plant used for making tapa is the very same brought to Japan centuries ago and now used for papermaking—namely *kōzo*—he experimented and produced his own variety of the beaten bark cloth.

### *Kamiko Manufacture*

HISTORICALLY the names of papers used to make *kamiko* differed with the various districts in which it was made, but essentially the paper was the same—pure *kōzo* paper, fairly thick and of the best quality to endure repeated rubbing and crumpling. In Iyo (Ehime) it was called *senka*; in Hitachi (Ibaragi), *Nishinouchi*. The *kyōsei-shi*, literally "strongly made paper," made in Shiroishi today is a descendant of the traditional *kamiko* paper. The paper is specially scooped and "stroked" on the mold to produce an extremely strong sheet. The papermaker moves his screen from side to side about as frequently and vigorously as he moves it to and fro. This binds the fibers in such a way that they cross at right angles to each other, producing a sheet of paper that is difficult to tear from any direction.

The old way of making the common type of *kamiko* entailed first coating thick *kōzo* sheets with *shibu*, an astringent juice squeezed from unripe persimmons. The *shibu* turned the paper a rust brown color; this was considered less elegant than the white, *konnyaku* treated *kamiko*, which could be dyed. The paper was grasped between the hands and rubbed around a smooth wooden vertical post; or it was held with the fingers of one hand and rubbed across the back of that hand with the palm of the other. The paper never rubbed against itself. This massaging tenderized the fibers and gave them some elasticity to the point where the paper was soft and limp like cloth. Then the *kamiko* maker would place the crumpled sheets between his mattress and the floor and sleep on it for a night. This added more moisture and, with minimum labor, a little more rubbing took place as well.

In Kurodani today a good *kamiko* is made in a manner similar to the traditional white variety worn by samurai long ago. The paper is heavily coated with a solution of *konnyaku*, a type of arum root. Then the four corners of the paper are folded into the center, and gently crumpled into a ball. After a while the sheet is unfolded, coated again with some solution, refolded and wrinkled again. This process is repeated until the paper is properly supple. Then the sheets are dried flat.

### *Shifu Manufacture*

*Shifu* is made from the best unbleached *kōzo*, made in thin sheets and stroked only to and fro on the mold. By this method all the fibers line up parallel to the width edges of the sheet, making it doubly strong and resistant to tearing against the width, i.e. lengthwise, but comparatively easy to tear *along* the width. Often a large sheet is cut into four equal pieces, each about 38 by 53 centimeters. Approximately one hundred cut sheets of *shifu-gami* (*shifu* paper) are used to form the weft for a kimono.

The sheets of paper are cut across the width into strips about 1.5 centimeters wide, though this width may vary, with a 2-centimeter margin. The cutting is quite difficult and delicate work, especially for the finer weaves of *shifu*. By the Shiroishi method, the sheets require some moisture at this point and so are pressed between dampened straw mats or cloths. Each damp sheet is then placed upon a flat, porous stone and carefully rolled back and forth with the palm along with a little slaked lime. This twists the paper into strands of "thread." How well the rolling is performed greatly affects the quality of the finished *shifu*. Then every other strand is cut or torn through the top and bottom margins and twisted into a thread between the thumb and index finger until one continuous thread is formed. This twisting of the margin pieces gives *shifu* its characteristic slub. Each sheet of paper produces a thread about one hundred meters long. Finally a collection of threads are twisted together on a simple yarn stretcher wheel. In Tamba, Mrs. Kawaguchi merely twists the paper by hand as she feeds it with slightly dampened fingers through her spinning wheel.

The dyeing of *shifu* can be carried out at any of several stages—when it is still paper sheets, before it is made into threads, after the first twisting, after the spinning or even after it is woven. The dyes used are natural dyes, often the same ones used to dye cloth.

In the finishing of *shifu*, the ends of the woven piece are sewn together to make a ring of material. The piece is immersed in hot water and crumpled by hand to help prevent shrinkage. The *shifu* is then boiled in straw ash lye, rinsed thoroughly in a running stream, and dried in the sun. The rinsing and drying are repeated about thirty times for undyed shifu and fifty times for dyed material. With each drying the sun bleaches the white *shifu* whiter and makes the dyed *shifu* colors more vivid. Until this finishing work is completed, *shifu* lacks those essential qualities for which it is famous, such as good perspiration absorbency, softness, and strength in water—although *shifu* cannot stand rough rubbing in water.

Thanks to the efforts of Mrs. Kawaguchi and also the Shiroishi group, *shifu* is enjoying a small revival in Japan, since the higher standard of living has now inspired a sentimental longing for objects of an earlier era, an era of great resourcefulness and closeness to the earth. *Shifu* is used today in making small bags, purses and wallets, underrobes and kimono linings, summer obis and kimonos, tea ceremony cloths, and many other personal and decorative items.

*Evaluating Washi*

THE washi produced in Japan today is on the whole excellent paper, extremely durable and beautiful and capable of astounding wear. The best of it, and this is particularly true of *kōzo* paper, increases in weight, strength, and whiteness over time. Although the washi produced by a very few papermakers in Japan has been laboratory tested and ascertained to be able to endure over a century, the rigidly traditional techniques used to produce it are all too quickly disappearing from rural papermaking. Not all washi is good washi, in short. A large percentage of that imported by the West is rather mediocre in quality, so much so that Westerners who choose from and use these papers would do well to learn how best to judge their merits and limitations.

Above all else, in judging and selecting papers it is an elementary yet often overlooked fact that a paper should be suited to its use. The most expensive, quality paper, then, is not necessarily the right choice for every project. Sometimes the visual qualities and appeal are the most important considerations; at other times qualities of strength and durability or a paper's capacity to avoid shrinkage are of paramount concern. Certainly it is a great advantage when working with papers to know well what is required of them for a particular purpose. Their selection involves being familiar with many different varieties and training the eye, ear, and fingers to recognize their many properties and differences before choosing among them.

*Observing Paper*

Chances are that a Westerner's first appreciation of Japanese handmade papers will be the attraction of the designs formed into them—the imbedded leaves or flowers or the wonderful texture of pieces of bark floating on the sheet. Well and good. But eventually many come to the point where their needs for a certain kind of paper become specific and it is essential to know and to be able to judge the real capabilities, limitations, and qualities of washi.

There are many ways to learn about washi. People who deal in and must choose among papers test the papers by feeling them, tearing them, smelling them, and listening to their rattle. Beyond this, one must rely on experience. I still think that the very best way to learn to judge papers is to approach them directly. *Use them.* As the Japanese say, to know the bamboo one must learn from the bamboo itself. To know paper, learn from the paper.

*Surface*

The visual properties of paper tell a lot about its quality. If possible, look at paper under indirect sunlight, such as by northern light coming through a window or by strong sun shining through a washi shade, the way many papermakers in Japan inspect their own sheets. Hold the sheet up to the light. Is it evenly formed, or are there thin spots? Are some areas obviously thicker than others? Does the sheet contain pieces of dirt or other impurities? Does it bear rough, unbeaten fiber? Is it clean? Has it been over-bleached? If dyed, is the color even and well absorbed by the fibers throughout? If it is *kōzo* paper, do the fibers seem to whirl in longish threads like fluffy clouds? Is the top surface of *kōzo* paper slightly furry, with its fibers able to be peeled off as it is rubbed? Does the paper have a fresh, crisp feel, or is it limp? In general, is it pleasing to the eye?

The upper and under surfaces of washi often differ more than they do in Western machine-type papers. Rub the paper lightly, holding it between the thumb and fore-finger, feeling simultaneously both surfaces. The fluffier, yet smoother side is the face, known in Japanese as the *omote*. The grainier, more bumpy surface is the *ura*, the back, which is the side on top when the sheet is formed and where the sediments accumulate. Although there is no hard rule, it is usually the face, the *omote*, which accepts the painting, writing, or print impressions. Sometimes the difference between face and back is negligible and quite difficult to distinguish, depending on the type of paper and its method of manufacture.

Paper has a definite grain, which is determined by the movement of the pulp solution as it is tossed back and forth and from side to side on the screen. The fibers naturally tend to line up in the direction in which they are moved; if they are tossed parallel to the long axis of the screen—that is, left to right and right to left—the grain will be "vertical." Fibers tossed toward and away from the papermaker, parallel to the short side (width) of the screen, form paper of "horizontal" grain. This is a common method, but generally the Japanese papermaker moves the screen in both directions alternately. Thus he interlocks and crosses the fibers so that it becomes difficult to tear the paper from either direction. Papers with "horizontal" grains are sometimes made for such purposes as *shifu* (paper "yarn" for weaving). *Shifu* paper must be as strong as possible in one direction, yet easy to tear in the opposite direction. Therefore the paper is generally made with a horizontal grain, the papermaker tossing the fibers to and fro on the screen. Imagine the *shifu* papermaker holding his mold just as a player faces the piano—the cracks between the keys line up in the same direction as fibers line up "horizontally" in the mold.

Grains can often be determined by holding the sheet up to the light and observing how its fibers line up; or observe, if the paper is thin, how the chain lines fall—the numerous and thin watermarks represent vertical bamboo splints of the screen, while the thicker more widely spaced marks are the horizontal threads (see Pl. 185). If neither of these methods clearly indicates paper grain, tear the paper at the corner in both directions. Only a small tear is necessary. If one side tears more easily and cleanly than the other, the direction of that tear is *with* the grain of the paper. There is an idiom in Japanese, *yokogami-yaburi*, which literally translated means to tear paper sideways against the grain; as it is idiomatically used, however, it means perversity or pig-headedness.

*Strength*

Of course a paper should be strong enough for its intended use. If a snap judgment must be made, long-fibered, unbleached papers are generally stronger by far than short-fibered, bleached varieties, and thick papers are usually but by no means always stronger than thin ones. Strong paper has a certain quality of substantiality about it; it is crisp rather than limp. If the paper's deckled edges are uncut, called *mimitsuki*—"with ears attached"—try to draw out stray fibers, noting if they are long, tenacious, and fibrous. This should indicate a strong paper. One of the best tests for strength is to tear the sheet horizontally and vertically. The more difficult it is to tear, the stronger the sheet, especially if it resists well from both directions. Good-quality *kōzo* paper looks and feels furry along a torn edge or rubbed surface.

Strong paper high in tensile and compression strengths can endure numerous foldings either with or against the grain. It can also withstand rough, constant handling before it shows signs of fatigue. Bend it, tear it, fold it, handle it, and use it—these are the ultimate tests for paper strength, although they may be conducted at the expense of a good sheet. If the sheet cannot be spared, test only a small area along a corner.

*Sound*

"Rattle" refers to the sound a sheet of paper makes when shaken. The ability to discern good rattle comes only from experience, from familiarity. Naturally papers that differ in type and thickness differ in sound as well. Quality washi has a sound to it that is very difficult to describe in words, but it is a clear, rich, crisp sound, more dry than damp, and rather crackling. The Japanese describe it as *pahsh-pahsh-pahsh*. The rattle of fairly thick paper, or quality washi, is deep and resounding. The noise it produces seems to have direction to it. If you listen with concentration when you shake good paper, you can actually hear a sound that gives the impression of quality and substantiality.

*Smell*

Papers have the most smell right after manufacture, and hopefully it is a fresh vegetable or plant smell rather than the odor of bleaching agents or alkaline chemicals, or of vegetable fiber starting to decay. Paper made in hot, humid climates or where the fiber was processed so slowly that putrefaction had begun to set in may have a somewhat sour odor, which is undesirable. In time, most paper odors become negligible. "Living National Treasure" Ichibei Iwano boasts that his paper has no odor at all, or at least not a distinct one.

NOTES

half-title: Tokutarō Hanada and Kiyofusa Narita, *Kami-suki uta*, 2 vols. (Tokyo: Seishi Kinenkan, 1951). (group trans.)

1. Bunshō Jugaku, *Paper-Making by Hand in Japan* (Tokyo: Meiji Shobō, 1959).

2. Seichiku Kimura, *Shinsen shikan*, orig. ed. 1777, reprint (Tokyo: Kami no Hakubutsukan, 1977). (group trans.)

*Dying things are kept alive by thinking of them. So it is with washi.*

Bunshō Jugaku

# The Future of Washi

The American papermaker and paper scholar Dard Hunter wrote confidently in 1936 that he felt, unlike European hand papermaking, hand papermaking in Japan would "go on forever"—not only because washi has characteristics that cannot be duplicated well by machines, but also because Japanese materials are plentiful and Japanese labor cheap. How quickly events have changed what that American paper authority took for granted as unchangeable! It is still true that machines cannot come near to duplicating the warm and beautiful character of washi—but today raw materials are scarce, and Japan's labor no longer the "bargain" of the past. These changes, compounded with other problems, now threaten the survival of handmade paper in Japan.

What sustained the production of handmade paper for over a thousand years, and what has been threatening its existence for the past hundred? At the close of the Russo-Japanese War (1905), Japan plunged headlong into industrialization. The mechanization of paper techniques and eventual seizure of the industry's wealth by pulp paper production started the chain of toppling events leading to hand production's decline. Mass production by mechanization created great quantities of paper at very cheap prices, something the hand paper mills tried to compete with but could not. Thereafter, but for World War II, the production of washi came to be in inverse proportion to the per capita consumption of machine-made paper—the former declined, and the latter swelled to astounding proportions.

This mass production of cheaper paper cornered the market, forcing many hand papermakers out of business and creating a lack of successors—thereby, in turn, raising the price of handmade paper. Many washi craftsmen attempted to reduce the price of their paper by introducing some mechanization into their methods and adding pulp to the pure fiber, not to mention other adulterations of the traditional way. Some succeeded, and there are still papermakers producing overbleached sheets one by one in this semimechanized manner. But such paper was and is inferior to the quality paper they had been taught to make, and most papermakers found such methods self-defeating morally as well as economically.

The hand papermakers also found themselves for the first time in need of good leadership and organization like the guilds of old—but had none. Paper craftsmen needed respect for themselves and their craft, and new roles for washi. The old ways of life in which washi had once played an integral part were dying out. The demand for washi decreased dramatically; yearly an increasing number of hand papermakers stopped practicing their craft. In 1973 the formal figure for the number of papermaking houses in all of Japan was 850, although according to Makoto Yagihashi of the Bunkachō (the governmental Agency for Cultural Affairs), the real figure is closer to 750, with more houses stopping production every year. In 1978, the number was 650.

Today washi production seems beset with problems at every turn. In one respect it is surprising that handmade paper has survived at all, and especially surprising that in some remote areas it survives in all the excellence and vigor of ancient days. Some of the more weighty problems that trouble the future of washi and some of its more encouraging influences will be discussed in this chapter.

*Materials*

PAPER is, of course, only as good as the fiber from which it is made. The present-day problem of a lack of fiber and its increasing price is therefore a crucial one to Japanese papermakers.

Traditionally, papermakers depended on the vast available resources of native *kōzo*, *gampi*, and *mitsumata* growing wild in the surrounding countryside in addition to the cultivated forms of these plants. But over the ten centuries of papermaking in Japan, papermakers have stripped the land heavily of its natural resources and have failed in recent years to replace the cultivated trees as fast as they have exhausted them. In the case of the *gampi* plant, which has eluded domestication, this in all likelihood spells imminent extinction. During and immediately following World War II, when Japanese federal laws forbade the cultivation of nearly all but food crops, the domestication of *kōzo* and *mitsumata* came to a near standstill. Before the war some eighteen million kilograms of *kōzo* were harvested a year, but by the early fifties, production was still not half that amount.

Although *mitsumata* is grown now in fair quantities, the best plants are reserved for the government treasury, where the fiber is used in making paper bank notes, and papermakers must make do with what remains. *Gampi*, which cannot be cultivated in quantity, is becoming extremely rare—and *gampi* paper accordingly expensive. In recent years some papermakers have begun importing *kōzo* from Korea and *gampi* from Korea and China to fill the need, but all deplore its marked inferiority to Japanese fiber.

Today few papermakers can afford to spend the time and labor required to raise and harvest all their own fiber—and those who are experts in this field are steadily dying off. Instead a system has arisen whereby farmers in outlying areas specialize in the cultivation and harvesting of paper fiber trees, trucking the steamed-off bark directly to the papermakers as it is ordered.

Although such a system permits papermakers to choose the variety and quality of fiber as their pocketbook permits from what is on the market, the system has serious limitations. Since hand papermaking in general has been on the decline, less *kōzo* and *mitsumata* is cultivated, and prices have suffered a fantastic rate of inflation. Between 1969 and 1970 the price of domestic *kōzo* doubled; between 1970 and 1972 it doubled again, the price of black bark soaring from 5,000 yen (15 dollars) per 10 *kan* (37.5 kilograms) to 10,500 yen (31.50 dollars); in 1978, the price in Kōchi Prefecture was 50,000 yen (272 dollars) and had risen to 80,000 yen (364 dollars) in some areas.

Such increases have made unexpected demands on the papermaker, who has always spent a minimum on materials—by borrowing from nature and paying with his own labor. Although the prepared bark represents time saved for the craftsman, allowing him to devote more hours to actually making paper, the problem is one of the modern inflationary economy putting excess burdens on a cottage industry. It simply cannot function as it used to function. The problems precipitated by increased fiber prices have forced many papermakers out of business. Those who continue to make paper in the traditional way have necessarily raised their prices for their paper, to the protest of buyers.

At present the Japanese government has been granting funds to papermakers for the raising of their own raw materials in an attempt to meet the problems of severe shortage and inflated prices. This has undoubtedly helped ease the situation; however, much more must be accomplished than individual papermakers raising only enough fiber for their immediate use, which is hardly different from what was done in the past. Papermakers are beginning to realize the need for organization among themselves, which would allow far-reaching plans for fiber cultivation to anticipate the needs of the future.

### *Successors*

A SMATTERING of international paper experts have from time to time remarked that in this labor-saving age, there is no way to make good paper by using labor-saving techniques. Its beauty, it seems, demands no less than making each individual sheet with a great deal of labor and a fair share of pain. Probably this problem more than any other is threatening washi with extinction. And it is for this reason the great papermaker Ichibei Iwano refused to teach the craft to anyone other than his immediate family. "The work is too arduous," he insisted. "The young can't take it." I have visited not a few villages where no one under fifty is making paper. What will happen when these elder craftsmen die?

It is a particularly difficult problem, for all the hard, hand labor that has given washi its warm, hand-reared character is now what the paper craft's would-be successors refuse to endure. No one can blame the youth—it is not that they are lazy or incapable of hard work. Rather it is that papermaking as it is today calls for too much sacrifice for too few returns in a swiftly changing and increasingly materialistic culture.

To outside observers, the work often appears idyllic, but the children of papermakers know otherwise. It is almost an adage in Japan that the work of making paper by hand is hard and unprofitable. As happened to Yashiro near Kurodani, paper towns are turning into ghost towns (see p. 131) all over Japan. The young seek salaried jobs in the cities, so that they might have the comforts and leisure time of the forty-five-hour work week that their parents never had. Making paper is hardly a romantic way of life to these young people.

Quite naturally, the older generations are very concerned about the future of washi. They deeply regret seeing the accumulated knowledge and skills that were passed down through their families for centuries die with them. Today such craftsmen who know every aspect of papermaking are as rare as the paper dolls once thrown upon the water.

It is clear that the number of successors will increase only if papermakers of this generation make their craft more desirable to their children. Only then can washi remain in the realm of folkcraft. This means creating a way of work and of life that is more satisfying than grueling. It also means instilling in the young a love of paper and its manufacture—a hard lesson, for the distractions and novelty of city life are great temptations to Japanese youth and tend to abort the growth of feeling and aesthetic depth that village crafts could offer them. Moreover, the children of papermakers need some assurance that papermaking as a livelihood will survive, for the young need to believe in the continuity of their work. It is possible. I heard of one young man who, disillusioned with city life, returned to his hometown of Kurodani to continue the paper craft he had learned when younger. This is an isolated case, but more may act as he if they see their elders believing in washi, improving the work conditions, and doing all they can to help it endure beyond their own lives in its pure, traditional way.

*Integrity and Technique*

IT IS in a sense ironic that the very thing that today threatens the future of washi is that which has kept it alive and pure in the past—poverty. "Poverty is Japan's greatest asset," said potter Kanjirō Kawai. Poverty in this sense means abundance of hard labor, natural methods, and an austere approach to life and work. It is just this affinity and intimacy of the craftsman to his materials, and the working of them within their natural environment, that has made washi so beautiful. But hand labor is strenuous, especially when large output is demanded, and craftsmen find it difficult to keep the spirit of honest labor fresh in the age of machines. The question is whether or not increasingly affluent Japan can continue to nurture such a craft, if papermakers must continue to endure such a rigorous existence and receive such small returns.

Bernard Leach wrote: "One secret of traditional handiwork is the absence of wasteful effort of mind as well as body."[1] This describes the Japanese hand craftsman, who has reached high levels of efficiency. Papermaker Eishirō Abe has remarked that, even though papermakers barely make a profit, they survive because they never waste any materials, and their technical skills are so developed that they rarely ruin a batch of paper. Indeed poverty has demanded it, for the difference between processing each season's fiber properly or ruining it was the difference between eating and starving. Therefore techniques that have been passed down over hundreds of years are more than just venerated forms—they are the pinnacle of economy, efficiency, and closeness to nature. "When we make paper," a group of papermakers wrote in *Tesuki washi*, "the life of each sheet is at stake."[2]

Economy of effort, efficiency of craft technique—these appear to be employed to their utmost while still retaining the human touch. Yet, all papermakers reiterate: we must reduce the price of washi. The big question is how can the price of washi be reduced without hurting its quality?

An unfortunate number of papermakers today have compromised their professional integrity in an attempt to lower the price of their paper—or to increase their margin of profit. It is not the case that the papermaker does not know how to make a good sheet of plain white paper, for the heritage is there for him to follow. Rather some have chosen out of laziness or mercenary interests or economic necessity to make bad paper. They short-cut on hand labor with machines and harsh chemicals and they adulterate the solution by adding wood pulp or inferior fibers. Often they throw gaudy pieces of fiber or other material into the vat or even use dye to hide what is basically poor paper.

The use of some machinery, the discriminate use of some pulp, or coloring, or inserting extraneous material into the solution or sheets are not in themselves necessarily adulterations—but the excessive use of any or all of these must be adulterations. My teacher, Seikichirō Gotō, impressed on me the precept that if a papermaker cannot make a good, simple plain sheet of pure-fibered paper, he is not worthy of the name of papermaker. Unfortunately, it is for the novelty of color or design that most people buy handmade paper, and so some makers exploit the interest in gimmickry. Sometimes papermakers have gone so far as to sell their inferior product under the name of another famous paper, as I recently witnessed with *Shiroishi-gami*. Sadly enough, such fraud only serves to undermine the excellent reputation of real *Shiroishi* paper.

When my teacher traveled over Japan in the fifties and sixties on pilgrimages to remote papermaking villages, he experienced many uplifting things—but also some disheartening ones. After visiting one village whose paper disappointed him greatly he wrote:

> What makes me wonder . . . is that despite the advantages of nature or environment and also the use of up-to-date equipment, the product [paper] is devoid of life. It has neither the beauty nor the strength that characterize Japanese paper. The expenses arising from large-scale equipment; the great effort to compete with machine-made paper; above all the ignorance of the good qualities of Japanese paper and the profit-first principle—these lead the maker to the production of lifeless paper . . . In order to maintain the production of good paper, the maker must be convinced of the importance and continuity of his work.[3]

The profit-seeking mind bends to compromise. Compromise has never produced good washi, nor any other fine handwork. One Japanese writer observed that "paper should be the most personal of all handcrafts." I think he is suggesting that of all the crafts, papermaking has to be especially wary of compromising methods and materials. One must not make paper more elegant or less excellent than paper should be—to make paper in the most conscientious and honest way is an end in itself.

Again, how can the constantly inflating price of washi be reduced without hurting its quality? Some feel this is only possible by cutting out part of the expensive hand labor, i.e., to mechanize some areas of the work. Of course in the long run this would help reduce the price of each sheet and alleviate problems of labor shortage, but at what expense to the quality of washi?

The question of mechanization in washi is difficult, but one might try looking at mechanization of the folk arts as Sōetsu Yanagi did: "The problem is not a matter of either hand or machine, but of utilizing both. We have yet to discover just what is suitable work for each."[4] Semimechanization of labor has already been put into effect by most papermakers today in varying ways and degrees, and some very good paper has been the result.

As might be expected, certain aspects of the work lend themselves more readily than others to the machine. One of the most universal applications of machinery is in beating the fiber, a very strenuous task when done by hand. Tadao Endō's hand-beating equipment in his Shiroishi workshop, for example, is on display but apparently not in frequent use. "But everyone machine-beats these days," he said half-apologetically. "They say it's just as good." And perhaps it is, because his paper is truly wonderful. One reason for Endō's success in marrying a quality product with market economics may be attributed to an interesting machine of his own invention, which expedites the stripping away of black bark. Here is an example of machinery efficiently replacing tedious hand work while maintaining the quality of the raw material.

In the 1930s the military government's Small Enterprises Committee made a nationwide investigation to determine how papermakers might further mechanize the craft without taking the life out of the paper as well. Some of their suggestions included use of a pressurized steamer for boiling the fiber, large screens and frames, metal indoor dryers in a cylindrical shape, and mechanical jack presses for removing water from the sheets. Their conclusion was that proper use of such devices actually improved the strength of each sheet. Today, free of political suppression, more conscientious papermakers have taken umbrage with these conclusions. Still, these machines are in widespread use. Such mechanized papermaking is called, ironically, *kairyō-suki*—"improved papermaking."

Perhaps mechanization in papermaking is inevitable, but one must be aware of its potential dangers. For one, as history has born out, whenever attempts have been made to mechanize a handcraft—whether for purposes of increased production or profit or whatever—the result has been cheap, low-grade products. This is what is happening; the many inferior kinds of washi now attest to it.

Undoubtedly some people feel that washi must make compromises, must employ some forms of mechanization in the future for the sake of keeping down the price. The trouble is that, in so doing, the life of the maker is removed more and more from each sheet, and the paper itself becomes lifeless. To me there is something timelessly beautiful about the traditional cycle of papermaking, replete with all the hard, hand labor. When I look at a sheet of purely made washi, for me all the work that has gone into it is not wasted. I appreciate it down to the last hand-beaten fiber—in fact perhaps I appreciate it *because* it is hand-beaten.

It seems that there is everything to be trusted in the traditional methods of manufacture, for they preserve forever a link between man, nature, and the paper he makes. Within the restraints of technique, the craftsman has both the freedom and the responsibility to express, not so much his individuality, but his own common and creating nature. Speed and volume and refinement have nothing to do with it. The value of the papermaker's work, I think, must be in his ability to create a bond between himself, nature, and his materials, and to maintain his integrity in all aspects of his work. Only with this understanding of work can real washi survive.

### *New Roles*

THE generations of middle-aged and older papermakers in Japan have experienced immense and shattering changes in their lifetimes. Modern industrialization has brought about totally new life-styles, ones in which washi plays a much less essential role. Traditionally every household required paper for two main purposes: writing and drawing and for use in *shōji* and *fusuma*. Therefore washi had to be strong, beautiful, relatively smooth and white, for the functions of paper were intimate and constituted established parts of peoples' lives in old Japan. Today machine-manufactured paper can fill such roles. And so the demand for handmade paper has been declining, although many Japanese express the feeling that, as washi disappears from their lives, some of life's warmth disappears with it. I have been told that as long as there are *shōji* there will be a demand for washi. I wish it to be true.

Papermaking in Japan has always been a folkcraft, but in recent years paper has begun to lose its folkcraft simplicity. Washi seems to be a little less approachable for the Japanese now than, say, thirty years ago. Due to its price and declining production, more and more it is becoming a self-conscious luxury item and less a daily necessity in the lives of ordinary people. This is a danger, for washi is no longer washi if it has lost its common touch. The price of an ordinary sheet is above what most people can pay to use it in ordinary ways. I have visited papermaking villages where the people themselves do not use the beautiful *shōji* paper of their own manufacture but use machine-made paper instead.

As it becomes further divorced from the everyday life of the people, washi tends to become unhealthy. The multiplying "folkcraft" shops and department store corners throughout Japan sell craft objects—including washi—to an appreciative urban public. Yet the very existence of such shops implies an urban demand. The nostalgic longing for expensive, quaint rusticity favored in the city is quickly sensed by rural craftsmen, who must know their markets in order to make a living. Such influence is insidious, at least in part. At the same time, urban demand provides a challenge. Only a constant and familiar use of washi will preserve its robustness and its wholesome nature. The traditional strength, durability, and honest character will continue to be worked into it by the papermaker only insofar as these qualities are *needed* for its use.

Therefore the future of handmade paper, as well as its health, depends heavily on what its functions will be. The first step is to realize what washi is not: washi is not machine paper, nor should it try to compete as such. As many observers have noted, machine and handmade paper should develop along separate but parallel lines, each with its own distinct purposes and uses; for each has its own possibilities as well as its own limitations. The virtues of machine paper—cheapness and the ability to be mass-produced—should not be washi's virtues. Machine paper is excellent for short-lived purposes; but where paper must be beautiful and durable, through time or wear, handmade paper is well suited. Machine paper and handmade paper can exist side by side in harmony and respect, for there are places for both in this world. But one should not try to become the other.

The potentials of Japanese handmade paper have yet to be fully realized and developed. One field that could use further development is the use of washi in hand printing. The mechanization of printing presses and of papermaking have gone hand in hand, the former developing increasingly more complex and innovative technical capabilities as paper has come to be smoother, more consistently even, of uniform weight and thickness, and available on continuous rolls. Printing on washi has its problems, as has printing on any handmade papers—they are better suited to the hand press than the web press, for the hand printer has more control over the irregularities present in each hand-formed sheet. In fact, the excellent printing qualities of washi have been lauded in the West by many printing experts. Washi has been used in the West since Rembrandt's time for sketches, woodblock prints, and etchings, and today in even more diverse ways involving our more advanced printing technology.

In other fields, the strength, beauty, and relative translucent qualities of washi can be used to advantage, for example, in lamp or window shades or interior lighting. Washi could also serve in photographic studios as warm reflectors of harsh light.

Retaining washi's vitality is largely in the hands of the creative talents of modern papermakers and designers. Paper craftsmen must develop a feel for the natural evolution of their craft in their environment and be ready to direct its course along those lines. They must find new functions that are true to people's needs, not extraneous functions. Yanagi said, "Apart from use and the people there is no meaning either in craftsmanship or its beauty."[5] And these functions must also be true to the nature of the paper itself.

To some extent papermakers are already using their creative imaginations in finding interesting ways to apply washi artistically. This is being done in Kurodani, in Yatsuo (Toyama Prefecture) under Keisuke Yoshida, and by Keisuke Serizawa's Tokyo studio, to name a few, where paper is dyed in various ways and formed into note cards, prints, book covers, and other items. Some of these items can be called true folk art in the sense of being usable objects that are honest, natural, and beautiful; others, luxurious craft work done in folk tradition; but unfortunately many also have a contrived feeling, where washi's natural beauty is covered up and where washi is used when other materials would serve as well. The real and important challenge, I feel, is to create uses that will retain the traditional flavor and pure feeling of handmade paper. Those will surely be roles that demand from washi all the strength, beauty, and durability required in past ages.

*Organization*

PAPERMAKERS in the past have rarely had the opportunity to communicate with one another on a nationwide basis; but today, with the news media, books, and television not infrequently focusing attention on the demise of this craft, papermakers have been coming together. Washi groups have had an acute need for communication, and now, too, organization, on a national level as well as on a local one. Washi needs leaders with energy and creative insight to point the craft toward true goals, for it is at a turning point in its evolution—it might as easily disappear as enjoy a great revival.

On the local level, groups must be organized to preserve traditional techniques and the flavor of the papers native to their areas. To some extent this is already being done, notably with such famous old papers as *Shiroishi-gami* in Miyagi Prefecture; *Izumo-shi* near Matsue, and *Sekishū-banshi* in Misumi, both in Shimane Prefecture; *hon Mino-shi* in Mino, Gifu Prefecture; and *Echizen hōsho* in Ōtaki, Fukui Prefecture. These are great papers of long tradition. The preservation groups that today keep their own heritages alive were all started by men of vision. Most have finally won government subsidization for their efforts, and it is hoped that in the future similar groups will arise to preserve the great papers native to their regions before it is too late.

In centuries past, when Japan was a feudal nation of self-sustaining provinces, papermaking was an important craft encouraged and patronized by the various daimyo. Although papers were traded extensively from one region of the nation to another, the separate feudal economic units and the guilds within them protected the crafts. Then the growth of urban areas helped the expansion of the crafts within the small village economies by creating greater markets for paper. These local economies had no need for organization at broader levels.

Today presents another situation altogether. The strengths of local leadership and organization have been responsible for the few successful paper mills now in operation. But I feel the individual paper communities are sorely in need of national organization to help foundering mills adapt themselves to a rapidly changing world economic system and to reinstate high standards in the hand paper industry. An all-Japan committee of this type already exists, but as yet it is too young and powerless to be an effective aid to the industry. The groundwork has been built—the group now depends on strong, vigorous, and dedicated leadership.

Here are several important functions of this committee that I feel need particular emphasis and attention:

First, papermakers need the establishment of standards and goals for all member papermakers, much like the role of the guilds of old, to preserve and protect the most excellent traditional qualities of washi. Perhaps only in this way will poor craftsmanship and the immoderate use of inferior materials be effectively discouraged.

Second, the committee could function as a seat of cooperation for the production and distribution of materials. At present papermakers are dependent on outside labor to fill their needs and are victimized by erratic production of materials at inflated prices. Cooperative efforts here could well mean cheaper and higher quality raw materials for all papermakers, and cheaper and better paper at the marketplace.

Third, cooperation in the manufacture of tools, machines, and equipment would be a big advantage for all paper craftsmen. In earlier times, papermakers made all their own equipment, but recently they have depended upon professional carpenters—who usually have no connection with paper—for their more complex equipment. As can be expected, such equipment is often faulty or inefficient. Those craftsmen who are also good equipment technicians are extremely scarce; when they die, they will take with them their expertise.

A fourth function would be as a cooperative acting as a wholesale distributor of manufactured paper. Local cooperative wholesalers exist now in a few isolated cases—but a cooperative of all papermakers would aid all without favor and help reduce the price of paper by cutting out scores of middlemen. Well-applied, creative imagination would be profitable here.

Fifth, the industry requires the creation of new roles for washi, as discussed earlier in this chapter.

Last, the committee must raise money for the preservation of fine existing papermaking, for the revitalization of fine old papers, and for the promotion of the public's appreciation and knowledge of washi.

Thanks to the persevering efforts of the Zenkoku Tesuki Washi Jittaichōsa (All-Japan Handmade Paper Fact-Finding Committee), the Japanese government has begun issuing grants of money to make possible the continued production of certain papers. This is a remarkable first step. But it is up to the papermakers themselves to make it all work—to open up youth to the beauties of this craft, to make it work economically as a profession, and to demonstrate more power of organization, more initiative, and more creative energies than they have yet displayed in order to build a nationwide cooperative group into paper's influential arm.

*Patrons and Promoters*

JAPAN has always been a culture-conscious nation, valuing highly her great artists and art treasures. The practice of designating certain living artists and craftsmen as "Living National Treasures" (see p. 30) began after World War II. Artists so honored receive some monetary grants for the protection and preservation of their craft. In 1968 the world of paper was honored when two papermakers were named "Living National Treasures": Ichibei Iwano of Ōtaki-mura, Fukui Prefecture; and Eishirō Abe of Yakumo-mura, Shimane Prefecture, for their skills in making *hōsho* paper and *gampi* paper, respectively. As part of the award, each receives the equivalent of about $3,000 a year to teach disciples and preserve the special traditional methods of making their papers.

In 1963 the All-Japan Handmade Paper Fact-Finding Committee mentioned above set out on what became a four-year investigation of papermaking villages throughout the country. As a result of their findings and evaluations, in 1971 the government of Japan began giving monetary assistance to two washi cooperatives, the *Sekishū-banshi* Technological Group (*Sekishū-banshi no Gijutsu Shakai*) and the Group to Preserve *Hon*

*Mino-shi* (*Hon Mino-shi Hozonkai*). This decision on the part of the government to help support these papermakers and cooperatives was a great stimulus to the washi industry at large, for handmade paper was at last recognized to be an important cultural asset in danger of extinction and worthy of revival efforts.

Today there are several groups and individuals fighting to preserve washi from extinction. A few devoted lovers of washi, such as Bunshō Jugaku and the late Sōetsu Yanagi, have tremendously elevated the position of Japanese handmade paper and brought it to worldwide attention. Moreover, they have given inspiration and encouragement to hundreds of papermakers. Similarly, the Morita Japanese Paper Company of Kyoto has promoted and supported the work of hand papermakers since 1926. Its present managing director, Mr. Yasutaka Morita (son of the founder) is both friend and mentor to papermakers throughout the nation, and his love for and knowledge of this craft is profound.

But most of washi's preservers are the papermakers themselves. These paper craftsmen, largely simple men, wield what influence they can both on an individual level—by hard and earnest labor producing their papers—and through the power of their cooperatives. Their steadfast efforts have earned them some financial subsidization and have won some recognition, interest, and sympathy from the public, which is beginning to appreciate at the last hour the beauty and worth of this dying craft.

Certain names appear and reappear frequently when one talks of Japan's handmade paper. Even the great machine paper manufacturers know these makers' papers and respect them. They are simple papermakers, but men of foresight, industry, and skill responsible for the health and revival of papers native to their regions. Aside from Iwano and Abe, many more deserve attention: Hajime Nakamura of Kurodani, Kyoto Prefecture, director for many years of the Kurodani cooperative union, which has been a kind of model cooperative under study by other washi groups; Keisuke Yoshida of Yatsuo in Toyama Prefecture, whose mills produce not only fine papers but also beautiful stenciled paper designs; Chika Naruko of Ōtsu, Shiga Prefecture, and an Intangible Cultural Property of that prefecture, who makes one of the most excellent *gampi* papers in the country; the Konbu family of Yoshino, Nara Prefecture, who produce fine *urushi-koshi* and *tengujō*; Yasukazu Kubota of Misumi, Shimane Prefecture, who directs the production of the wonderful, traditional *Sekishū-banshi*, and Kiyokatsu Shimizu of the same town, who has developed his own special *gasenshi*; Kōzō Furuta of Warabi (Mino), Gifu Prefecture, who has revitalized and preserved the famous *hon Mino-shi*; Tadao Endō of Shiroishi, Miyagi Prefecture, a superior papermaker in every sense, and his fascinating colleague Nobumitsu Katakura, once baron of Shiroishi, who with his daughter and the Satō family have revived the noble art of *shifu*, woven paper textiles; Zenzō Oikawa of Iwate Prefecture; Seikichirō Gotō of Fujinomiya, Shizuoka Prefecture, a versatile paper artist as well as papermaker, writer of many books on papermaking, my teacher; and many more.

These people are just a sampling of the thousands of country papermakers, some of whom I visited, most of whom I could not. They are often simple farmers making paper as a side industry in hundreds of small towns dotting the countryside in Shikoku, in Kyushu, Wakasa, in central and northern Honshu and the hills of Kinki, wherever the air gets cold and the water is good. These able craftsmen, often working as communities but also solitarily, devote a good part of their life to the manufacture of paper, producing it despite the great effort expended and despite the negligible monetary returns, for they love papermaking and they love paper.

*Zenkoku Tesuki Washi Rengōkai (All-Japan Handmade Paper League)*

THE handmade paper industry was one of many cottage industries restricted and conscribed by the Japanese government during the Second World War. Papermakers were forced to make paper only for military use, particularly for bullets and paper bombs. In 1939 the government created the Zenkoku Tesuki Washi Kōgyō Kumiai Renmei (All-Japan Handmade Paper Federation of Cooperative Unions), its purpose being to ration and distribute raw materials to the conscripted members.

In 1941 the group changed its name to the Japanese Handmade Paper Industrial Cooperative's Labor Union, and changed its function—to promote communication among the members, to increase their profits, and to develop the union's influence. Wartime government intervention had hurt papermakers, and the military draft had stripped the craft of its manpower. A more Western manner of living that followed the war also led to a decrease in the use of handmade paper. Eventually the group lost its influence with hand papermakers, but was taken up by machine papermakers and is now their national organization.

The present Zenkoku Tesuki Washi Rengōkai was established in October of 1963 in Matsue, Shimane Prefecture, by papermakers and other culturally concerned people who love washi and feared for its future. Its main purpose is to preserve the old methods for making paper—"protecting it from new approaches." The group is still young, but has already wielded its influence by helping persuade federal cultural agencies to aid and protect some great traditional papers. The group publishes its own information bulletin.

*Bunshō Jugaku*

Now in his seventies, Dr. Bunshō Jugaku is presently a professor of English literature at Kobe University. His interest in Japanese paper began with his interest in old English manuscripts; trying to gauge the age of certain old English texts, he began studying the paper on which they were written and the watermarks laid into the paper for clues of age and origin. He found his study of paper immensely absorbing but also continually repressed and frustrated by his country's militaristic nationalism, which deplored all things Western in prewar Japan. So upon the suggestion of a teacher whom he respected highly, Dr. Jugaku began the study of his own country's handmade paper. Today he is one of Japan's foremost experts on the subject.

Dr. Jugaku's deep love and reverence for washi is very apparent. He has been a tireless writer, lecturer, and essayist, inspiring interest in and respect for handmade paper and the craft of papermaking. He has been a friend to papermakers, offering great encouragement through difficult years. The townspeople of Kurodani hold him in particular esteem, for they feel he has played a leading part in the revival of papermaking there. Besides essays and articles in periodicals, Dr. Jugaku has written the following books: *Washi fudoki* ("Japanese Handmade Paper Seen Historically and Topographically"), 1941; *Paper-Making by Hand in Japan*—in English, 1942, and in Japanese, 1944; *Kamisuki-mura tabinikki* ("Diaries of Pilgrimages to Papermaking Villages"), 1943, with his wife.

*Sōetsu Yanagi and The Japan Folk Art Society (Nihon Mingei Kyōkai)*

THE great inspiration behind the interest in Japanese folk art in recent times was one man, the late Sōetsu Yanagi. In 1926 or thereabouts on a historic night on Mt. Kōya, Yanagi founded the Japan Folk Art Society in conjunction with potters Shōji Hamada and Kanjirō Kawai. It was Yanagi who coined the Japanese word for folk art, *mingei*, by condensing the clumsy phrase *minshū no kōgei*, combining the word-roots *min* (people, folk) and *gei* (art or craft). Due to Yanagi's ceaseless efforts throughout his lifetime, there is an increasing appreciation and understanding of Japanese native folk arts among the Japanese as well as abroad, and the term *mingei* is now not only a common Japanese word but is even becoming familiar in craft circles abroad. In 1936 Yanagi established Tokyo's Nihon Mingeikan, the Japan Folkcraft Musuem, which exhibits with refined taste and simplicity examples of old and contemporary folkcrafts.

Yanagi loved washi very much. He wrote several beautiful essays on the subject, the most famous of which, "The Beauty of Washi," appears in excerpts throughout this book. Yanagi sought out, discovered, and encouraged many of whom are today well-known and respected papermakers and artists, all of whom have developed their own unique styles without losing touch with the heart of folk tradition or losing a sense of intimacy with their paper. Yanagi, in part, must be responsible for this.

Yanagi showed the world the humanity of simple things. For him, a folkcraft object gains vitality and beauty from its use in daily life, just as the user gains vitality and beauty in his life by using it. Actually this concept is very old and deeply rooted in Japanese culture, even if never so explicitly expressed. But Yanagi sensed how modern life was leading men astray from their artistic roots, and for those who would listen, he pointed the way. Yanagi's greatness is also manifested in the heritage of craftsmanship ideals he left behind him, which echo the spirit of Tea. He wrote:

> Man keeps demanding his own personal liberty, but at the same time he falls victim to new restrictions of his own making. This is why the folk art movement seeks to show the profundity within the world of beauty of a world without self. In this sense the movement in Japan is more than a mere industrial art movement, it is also significant, to a certain extent, as a new religion of beauty.[6]

Yanagi also sensed the end of the world of the purely traditional craftsman and tried to point a way for the artist-craftsman of the future. Judging from the work exhibited today in Japan, there is encouraging evidence that some emerging young Japanese artist-craftsmen are seeing with Yanagi's vision. Whether they have read and absorbed Yanagi's often difficult writings themselves or perceived on their own the nourishing value of traditions is unimportant. What is important is that the contemporary craft world is opening up to creative innovation based on natural laws and to a recognition of objects that leave behind clever design as an end in itself, objects that aspire toward an essential but more intangible quality of fine crafts—natural integrity.

Today the Folk Art Society asserts its influence in many directions in the promotion of the folk arts. The society publishes a small monthly magazine, *Mingei*, devoted to Japanese folk art. In addition, it collects and exhibits the best of ancient and modern folkcrafts at its museums in Tokyo, Osaka, and Okinawa. There are also numerous private folkcraft museums throughout the country, all of which took inspiration from the work of Yanagi. Major private museums are in Kurashiki, Matsumoto, and Tottori, and the number is constantly growing.

The second director of the Japan Folkcraft Museum, the late potter Shōji Hamada, took charge after Yanagi's death in 1961 and became head of the society in 1973.

English potter Bernard Leach and the American scholar Langdon Warner were associated with the society from the start, bringing washi as well as other Japanese folk arts to the attention of the West. The Japan Folkcraft Museum maintains a concerned interest in the future of washi—they promote it by displaying it, for Japanese handmade paper will endure, they believe, if people are reminded of it and from this begin to use it in their daily lives.

*In Conclusion*

IN THIS age of plenty, this age of disposability and carefully planned obsolescence, it seems that there is little we value. Partly this must be due to the rapid change and impermanence characteristic of our culture, partly to the availability of "plastic" goods at low prices. It is not until now, with the impact of the energy crisis and general shortage of food and goods, that we begin to look at all that we had formerly taken for granted—the devastating attitudes and wasteful habits created by our industrialized economy, and the extent to which we have exploited our earth and everything upon it.

What is this new ecological interest, really, but an attempt to glean values concerning life on this globe, to try to find out how best to use the land and life that grows upon it wisely and with profound respect. It is a view both spiritual and practical, and in the end I think we will discover that it is the only realistic view.

Among the goods of the world, paper symbolizes the most transient, the most expendable and disposable. But for the Japanese, handmade paper, though common, is something valued highly. Until the last century, when all paper was made by hand and was therefore scarce and precious, people used it with great thrift. The recycling of handmade paper has been a commonplace practice for centuries. Even today Japanese do not take washi for granted but use it, as one explained to me, *arigataku*—with appreciation, with gratitude.

Many people consider the high price of washi one of its greatest drawbacks, although in these times of rapid inflation, Japanese paper is still at a reasonable price. Rather than trying to lower the price of washi, we might try changing our attitudes toward its use, using it wisely and reverently, making the most of its beauty and durability. As Yanagi wrote, ". . . articles well made by hand, though expensive, can be enjoyed in homes for generations, and, this considered, they are not expensive at all. Is spendthrift replacement economical?"[7] By wastefulness and careless use of what we have, we do violence to the earth in small ways. If everything we owned and used was paid for with our own hand labor, we would want to make these items last, we would cherish them, take care of them and want to use them respectfully, keeping these things for our children and our children's children. Would that we knew how to use things such as paper as the Japanese do—"with gratitude."

But the age in which people made goods, often of beauty, with their own hands for their own or community use has perhaps passed forever. In the West, particularly, it is now the era of the individual artist rather than of the communal artisan, the age of the well-formulated, sophisticated, original personal statement rather than the common will stating universal, if unconsciously expressed, truths.

Consider these words:

> We now live in an age where it is difficult to work collectively, since each individual takes responsibility for his own livelihood and works according to this premise. Even though some folk crafts are accorded protection by being designated Intangible Cultural Properties, the more conscientiously an artisan does his work the harder his lot is. Such difficulties can only be overcome by the individual's willpower and abilities. When we consider that the best work today is being done by artists rather than craftsmen, we can see how difficult a true craftsman's work is.[8]

Truly the age of folkcraft the world once knew is gone, and the craft of making paper by hand in its best traditional sense has small chance of surviving. The age when this cottage industry engaged whole families and communities in a cooperative endeavor has fallen aside to make room for the age of the individual artist and artist-craftsman who has new roles and new responsibilities, who, as Bernard Leach wrote, "is no longer a villager, but the inheritor of all cultures, and his power of assimilation has to be proportionately far greater. In fact he has to be a creative artist."[9]

The original washi—the unaffected, unself-conscious stuff that served the most banal functions and endured—that is perhaps gone forever. Though on the brink of extinction, it may yet exist in a few remote papermaking villages where the softest, chastest white paper is still glued to *shōji* or used for writing a letter; or a torn piece of *chiri-gami* is used to wrap vegetables, or as a handkerchief, or twisted into string. In such a place people will never waste washi, for, having made it themselves, they are aware of its value. They respect and cherish paper and use it daily in the most prosaic and natural ways, with affection. These are people truly intimate with paper.

Will Japan's handmade paper survive? I believe it will—for at least several ages to come, although unfortunately it will probably not survive in its best and pure traditional form. As my teacher Mr. Gotō said, this craft is in a disturbed and unsettled state, but "the techniques of papermaking are deeply rooted . . . and are strong enough to last as long as the race lasts."[10] Perhaps that is reason enough. I prefer, however, to think as Ichibei Iwano did when he said that it is impossible for the machine to duplicate all the charms of handmade paper. In any age, he said, great things endure. Washi will survive.

NOTES

half-title: Bunshō Jugaku, "Fukkō," *Ginka* 9 (Spring, 1972): 44–47. (group trans.)

1. Bernard Leach, *A Potter's Book* (London: Faber & Faber, 1969).

2. Bunkachō, ed., *Tesuki washi: Echizen hōsho, Sekishū-banshi, hon Mino-gami* (Tokyo: Dai-ichi Hōki, 1971). (group trans.)

3. Seikichirō Gotō, *Nihon no kami (Japanese Hand-made Paper;* also titled *Japanese Paper and Papermaking*), Vol. II, Western Japan, trans. Iwao Matsuhara (Tokyo: Bijutsu Shuppansha, 1958–60).

4. Sōetsu Yanagi, *The Unknown Craftsman*, trans. Bernard Leach (Tokyo: Kodansha International, 1972).

5. *Ibid.*

6. Hugo Munsterberg, *The Folk Arts of Japan* (Tokyo and Rutland, Vt.: Tuttle, 1959).

7. Yanagi, *op. cit.*

8. Kageo Muraoka and Kichiemon Okamura, *Folk Arts and Crafts of Japan* (Tokyo and New York: Weatherhill/Heibonsha, 1973).

9. Leach, *op. cit.*

10. Gotō, *op. cit.*

# PLATES

28. *Kōzo*

29. *Gampi*

30. *Mitsumata*

31. *Tororo-aoi*

32–34. Harvesting *kōzo*, Kurodani, Kyoto Pref.

33

34

35–38. Steaming *mitsumata* and stripping bark, Izumo, Shimane Pref.

36

37

38

39–44. Steaming *mitsumata* and stripping bark, Yamanashi Pref.

40

41

42

43

44

45. Stripping *kōzo* bark, Fukuoka Pref.

46–52. Steaming *kōzo* and stripping bark, location unknown.

47

48

49

50

51

52

53. Drying *kōzo* bark, Nagano Pref.

# Soaking Black Bark

54–59. Kurodani, Kyoto Pref.

55

56

57

58

59

# Scraping Black Bark

60. Family work, location unknown.

61–63. Yoshino, Nara Pref., 1947.

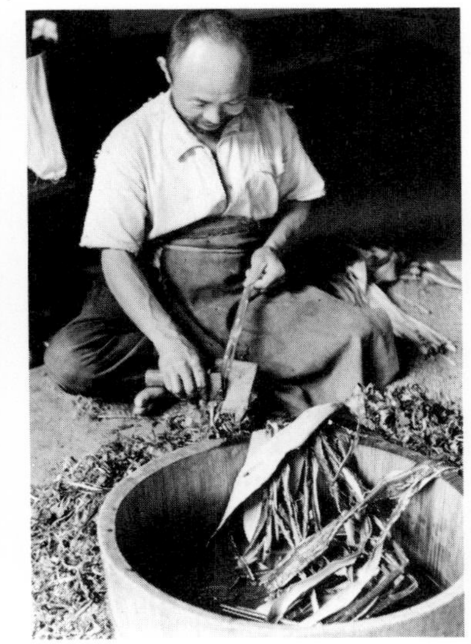

62

63

64

65

64–66. Yoshino, Nara Pref., 1974.

67. Kamikawasaki, Fukushima Pref.

68. Kurodani, Kyoto Pref.

69, 70. Yoshino, Nara Pref.

70

71. Kurodani, Kyoto Pref.

72, 73. Oguni, Niigata Pref.

73

74

74, 75. Machine scrapers, Shiroishi, Miyagi Pref.

76. Machine scraper, Gokayama, Toyama Pref.

# Boiling

77

77, 78. Yoshino, Nara Pref., 1947.

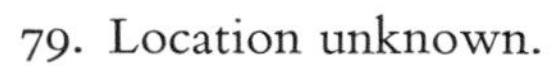

79. Location unknown.

80. Checking the bark, location unknown.

81–83. Kurodani, Kyoto Pref.

82

83

# Washing and Bleaching

84–87. Yoshino, Nara Pref.

85

86

87

88. Drying and sun bleaching, Yoshino, Nara Pref.

89, 90. Kurodani, Kyoto Pref.

90

91, 92. Warabi, Gifu Pref.

92

93–95. Ōtaki, Fukui Pref.

94

95

96–100. Kurodani, Kyoto Pref.

97

98

99

100

# Picking out the Dross

101–108. Warabi, Gifu Pref.

103

104

105

106

107

108

109. Kurodani, Kyoto Pref.

110. Hida-Kawai, Gifu Pref.

111. Kurodani, Kyoto Pref.

112. Nagano Pref.

113. Kurodani, Kyoto Pref.

114. Ōmi-Kiryū, Shiga Pref.

115, 116. Yoshino, Nara Pref., 1947.

116

117. Gifu Pref., 1959.

118. Tōno, Iwate Pref., 1950s.

119. Ōmi-Kiryū, Shiga Pref.

120. Karasuyama, Tochigi Pref.

121, 122. Ōtaki, Fukui Pref.

122

123, 124. Warabi, Gifu Pref.

124

125. Shiroishi, Miyagi Pref.

126. Akibatake, Gumma Pref.

127. Yoshino, Nara Pref., 1947.

128. Pulp mill, Iwahara, Kōchi Pref.

129. Beaters, Kurodani, Kyoto Pref.

130

130, 131. Beaters, Sakai, Nagano Pref.

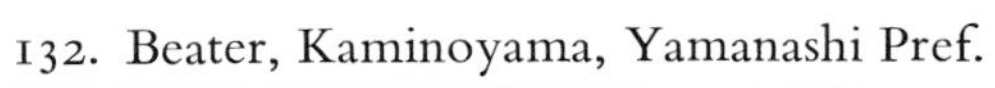

132. Beater, Kaminoyama, Yamanashi Pref.

133, 134. Hollanders, Kurodani, Kyoto Pref.

134

135, 136. Pulp vats, Kurodani, Kyoto Pref.

136

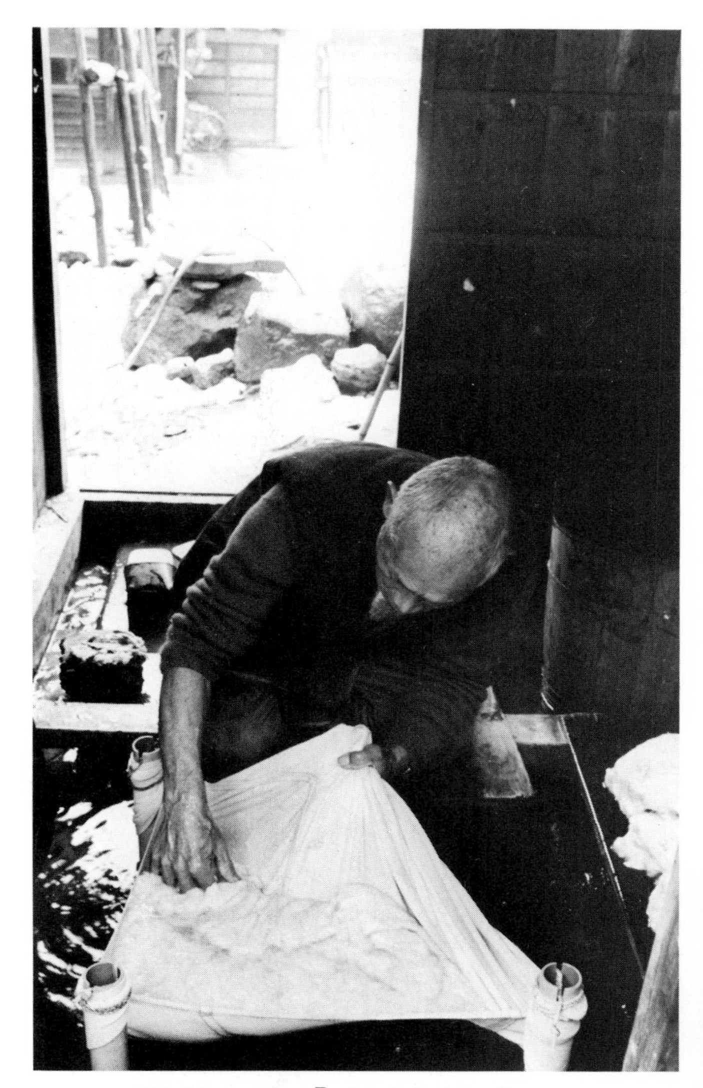

137, 138. Washing pulp, Ōtaki, Fukui Pref.

138

Mucilage Preparation

139. Pounding *tororo-aoi* roots, Misumi, Shimane Pref.

140. Pounding *tororo-aoi* roots, Kurodani, Kyoto Pref.

141. Pounding *tororo-aoi* roots, location unknown.

142. *Tororo* storage, Akibatake, Gumma Pref.

143. Scraping *nori-utusugi* bark, Yoshino, Nara Pref.

144. Crushing *tororo-aoi*, Kurashiki, Okayama Pref.

145. *Tororo* crusher, Kaminoyama, Yamagata Pref.

146

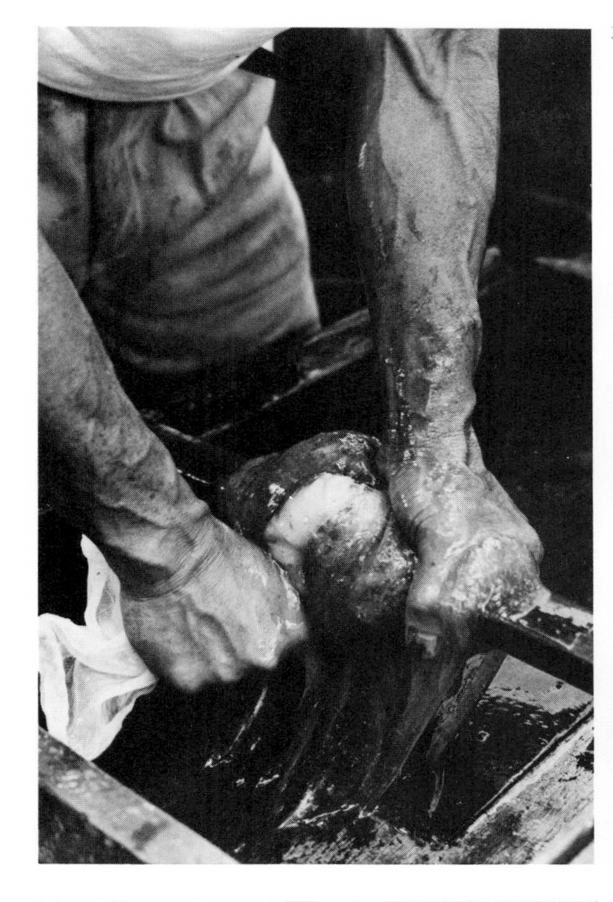

147

146–148. Filtering mucilage, Ino, Kōchi Pref.

149. Filtering mucilage into vat, Yoshino, Nara Pref.

150. Stirring pulp solution, Warabi, Gifu Pref.

151–153. Using vat agitator, Kurodani, Kyoto Pref.

152

153

154–158. Scooping, Kurodani, Kyoto Pref.

155

156

157

158

160

159–166. Scooping, Karasuyama, Tochigi Pref.

161

162

163

164

165

166

167. Yoshino, Nara Pref., 1947.

168. Yoshino, Nara Pref., 1973.

169, 170. Scooping, Yoshino, Nara Pref., 1947.

170

171–176. Scooping large sheets, Ōtaki, Fukui Pref.

177. Scooping postcards, Kurodani, Kyoto Pref.

178. Tossing pulp solution, Yoshino, Nara Pref.

179, 180. Tossing pulp solution, Ino, Kōchi Pref. and Warabi, Gifu Pref.

180

181. Eliminating dross, Warabi, Gifu Pref.

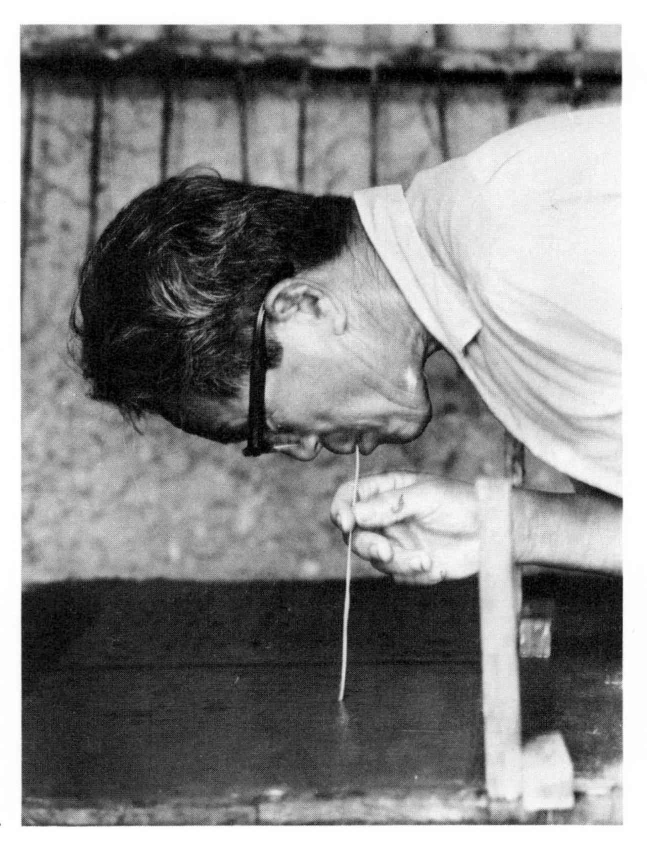

182. Removing air bubbles, Obara, Aichi Pref.

183–185. Lifting the screen from the mold, Karasuyama, Tochigi Pref.

184

185

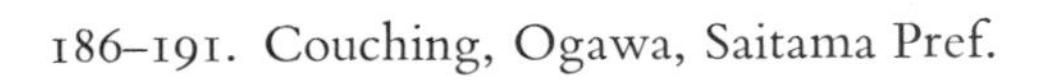

186–191. Couching, Ogawa, Saitama Pref.

192

192–194. Couching, Ōtaki, Fukui Pref.

194

195

195–198. Couching large sheets, Ōtaki, Fukui Pref.

197

198

199–201. Couching *Yoshino-gami* directly on drying boards, 1947 and 1973.

200

201

## Pressing

202. The simplest press, Yoshino, Nara Pref., 1947.

203. Takaoka, Kōchi Pref.

204. Weighted lever press, Yoshino, Nara Pref., 1947.

205. Weighted lever press, Ōmi-Kiryū, Shiga Pref.

206. Press weights, Ōtaki, Fukui Pref.

207. Water lever press, Takaoka, Kōchi Pref.

208. Screw press, Yoshino, Nara Pref.

209. Mechanical jack press, Karasuyama, Tochigi Pref.

210. Hydraulic jack press, Ogawa, Saitama Pref.

211, 212. Applying paper to drying boards, Kurodani, Kyoto Pref.

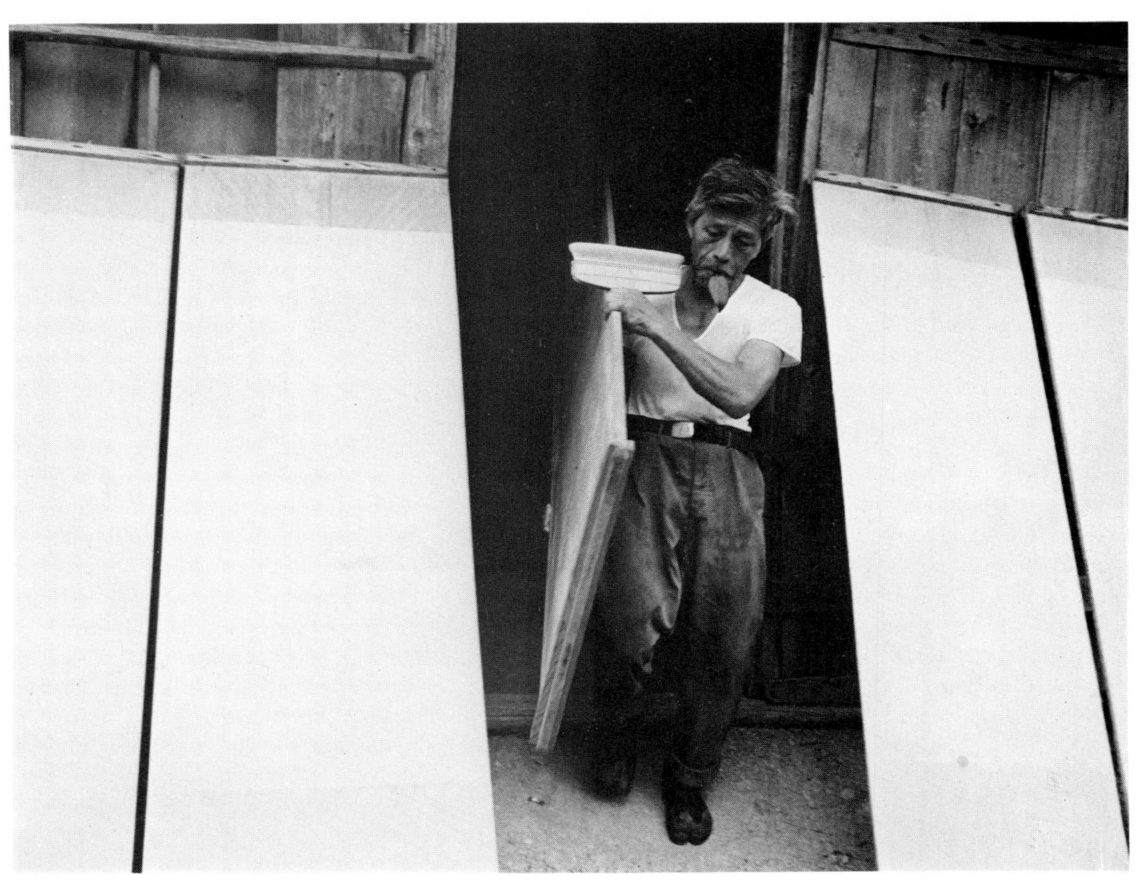

212

213–215. Setting out drying boards, Kurodani, Kyoto Pref.

214

215

216. Yoshino, Nara Pref., 1947.

217–220. Yoshino, Nara Pref., 1974.

218

219

220

221, 222. Cabinet dryer, Ōtaki, Fukui Pref.

222

223, 224. Iron sheet dryer, Kurodani, Kyoto Pref.

224

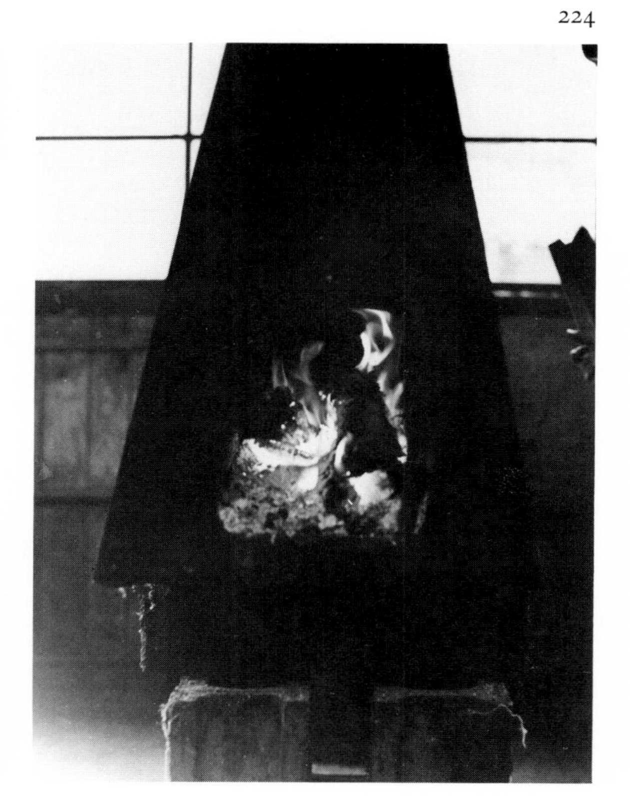

225. Trimming deckle edges, Yoshino, Nara Pref., 1947.

226. Storing finished sheets in snow, Oguni, Niigata Pref.

227 Nostalgia—taking paper to market, Yoshino, Nara Pref., 1947.

# Plate Captions

Note: Readers are generally about evenly divided in opinion concerning the amount of information that should be included on a photographic page. The balance of space, photograph sizes, and words is a delicate problem in making an attractive and informative layout. There are instances when verbal information in a photo layout interferes with the photographs' purely visual content. Such is the case in this book. The decision to separate captions and photographs was editorial, not the author's. Apology is made to all readers who find the necessity of page-turning an annoyance. (—ed.)

1. The area of Shikoku once known as Sanuki—modern Kagawa Prefecture—was and is famed for noodles, lacquerware, and kites. The latter developed an original and colorful style and remain popular today.

2, 3. See p. 61.

4, 5. Besides the work of professional craftsmen, umbrella-making in the Edo period even today has associations with attempts to make a living by impoverished, masterless samurai (*rōnin*). One, perhaps apocryphal, reason given for this is that an umbrella is raised above the head of the user. Thus, making such an object was not demeaning to the social position and pride of the samurai, no matter how poor. Today, the demand for handmade Japanese umbrellas is quickly diminishing, and the craft is dying; most of the umbrella-makers are elderly, though the skill continues in such traditional centers as Gifu and Kyoto. See p. 64.

6. See pp. 124, 200.

7–11. See p. 68.

12. The *koyori* hat, gourd-shaped water bottle, cups, and ladle probably date from the first half of the nineteenth century. The remaining objects are probably early twentieth century or just before. The hat type was worn by warriors or officials. This selection is a good cross-section of typical *koyori* objects, though the range of such forms was, of course, much greater. See p. 68. Private and Ōji Paper Museum collections.

13, 14. See p. 67.

15. The *dondo-yaki* observance occurs on the night of the fourteenth day or on the fifteenth day of the new year and involves the ritual burning of the pine and bamboo decorations used during the New Year celebration. It is the final purification of the New Year, representing the elimination of all vestiges of the previous year. Good-luck Daruma (p. 67) purchased the previous year are ritually burned also in certain parts of the country. New Daruma are then purchased.

16. Among the most famous of traditional papier-maché dolls and toys in Japan are those made in

Miharu, Fukushima Prefecture. These four excellent examples probably date from the late eighteenth century. The wooden molds used to form Miharu dolls are sometimes works of art in themselves. See p. 66.

17. The farmer's work coat from northern Japan is woven with a weft of white paper alternating with remnants of old cloth that have been indigo dyed, torn, and spun or rolled into yarn. The warp is cotton thread. The paper gives the coat warmth and a soft pliancy quite different from an all-cotton garment. The coat is forty to sixty years old.

The knotted warrior's undershirt dates from the early nineteenth century and was made in the Sendai area. The brown paper in the background is a large square of many layers coated with persimmon tannin (*shibu*) and probably used to hold silkworms. It is a patchwork of used papers, some of them containing poems.

18. The bolt of rough *shifu* is from Yamagata Prefecture, has a paper weft, and an unrefined hemp fiber warp. It is made of used letter or account book paper; traces of writing can be seen as flecks of ink on the paper yarn. Its use is conjectural. The age is about the same as the farmer's coat.

The fine cloth in the same photograph was used as a carrying or wrapping cloth (*furoshiki*) by the ladies of the household of the lord of the Mito domain, who was a member of the ruling Tokugawa family. It dates from the eighteenth or early nineteenth century. The warp is very fine cotton; the weft is equally fine *shifu*, and the weave is open and extremely delicate. The cloth is backed with silk, and the dyeing is subdued and elegant. The rough, rustic bolt of *shifu* and the aristocratic cloth make an interesting study in extremes. See pp. 56, 211. Undershirt and wrapping cloth from the Ōji Paper Museum collection.

19, 20. The *kamiko* kimono was coated with persimmon tannin (*shibu*) and wrinkled to give a rustic effect. The elegant silk trim accentuates the rusticity of the *shibu*-coated *kamiko*. This kimono was once owned and worn by a famous literati artist and poet of the late eighteenth and early nineteenth century. Private collection.

The *chabaori*-style short coat dates to about 1600. Considering it is over 350 years old, its unflawed condition is truly amazing, but the quality of the coat itself—its fine, stencil-dyed pattern, color, and design—is unsurpassed. Like the kimono, it has a thin padding of silk floss. Such a garment is an example of what can be done with fine-quality, strong, handmade paper, and, it is hoped, will be an inspiration to all textile and craftspeople who read this book.

The blue-and-green quilt cover is stencil-dyed in a pattern often seen on Japanese indigo cotton quilt covers, but the style is somewhat different. Its provenance is unknown; the date is probably late nineteenth century. Paper quilt covers are extremely rare—in fact, this piece was purchased in an antique shop, and its use is not completely certain. See pp. 53, 211. *Chabaori* and quilt cover from the Ōji Paper Museum collection.

21. This rare painting of a stencil-dyer at work is probably from a hand-scroll depicting various crafts and trades. Its date is uncertain; the most likely estimate is that it was painted around the late sixteenth or early seventeenth century. The cut-paper stencils (*kata-gami*) are clearly seen around the dyer, hanging from wooden supports. Such paintings are valuable records both of craft techniques and of textile designs of the time. Stencil dyeing today has become a popular hobby of urban house-

wives in Japan—the technique has not changed essentially from the one pictured. See p. 203. Private collection.

22. The five colored paper cushions are from the famous Yakushi-ji temple in Nara. Their date is uncertain, but they can perhaps be assigned to sometime in the eighteenth century. The lacquered hat, like the *koyori* example, is of a type used by officials or warriors. The boxes are monochrome examples of the usually more colorful *kinkarakawa* (see p. 141); the small, opened box is a tea container. *Shibu*-coated dustpans like the one pictured are still being made. See p. 69. Ōji Paper Museum collection.

23. Most of the colorful woodblock-printed books pictured are nineteenth century. Noteworthy exceptions are the books with blue and green covers printed with delicate gold and mica designs, which are early seventeenth century. The small volume showing what appear to be paint samples is a catalog of kimono color combinations. Private collections.

24, 25. See p. 62. Ōji Paper Museum collection.

26. Until about 1970 or so, *shōji* repaired with cherry blossom and flower-shaped patches could be seen everywhere in Japan. Today, glass, aluminum window sashes, and an affluent economy have made this sight very rare indeed. Similarly, even cracked glass panes were repaired with strips and flower patches of washi until such a time as the panes could be replaced. It was hoped that a photograph of the latter could be included here. Countless cracked and patched glass panes were found, but they all were held together with the same stuff—cellophane tape.

27. Drying boards at Eishirō Abe's workshop, Izumo, Shimane Prefecture. See p. 137.

28. *Kōzo* is a general term for several species of paper mulberry, genus *Broussonetia*. When the plant is two to three years old and its stalks are about as thick as the ones pictured, the bark is suitable for making paper.

29. *Gampi* (several species of *Diplomorpha*, *Thymelaeaceae* family) is the most prized of the paper fibers, not only for its qualities of length, strength, and fineness of fiber, but also because it defies cultivation and has become scarce. Centuries of papermaking have greatly depleted this wild plant, and papermakers today fear that it may disappear.

30. *Mitsumata*, also a member of the *Thymelaeaceae* family, genus *Edgeworthia*, is considered the softest of the three main Japanese paper fibers. The word *mitsumata* means "three-pronged" or "-forked"; this branching structure can be clearly seen in the photograph. The plant flourishes in Shikoku as well as in the area around Mt. Fuji and in Gumma Prefecture.

31. *Tororo-aoi* (*Abelmoschus manihot*, Medicus) roots are harvested in early winter before the first frost. Because of its lovely flowers, the plant is sometimes seen decorating gardens.

### Harvesting and Steaming

32. The second *kōzo* harvest at Kurodani occurs in February or March, yielding "spring bark." The stalks are cut about six inches above the root; the plant produces new shoots annually, but becomes unusable for paper between its tenth and twentieth year. Kurodani, 1972.

33, 34. The Kurodani *kōzo* farmer ties the shoots into bundles of just manageable size and weight—

about twenty kilograms each—and spears a bundle onto either end of a carrying pole. Using a walking stick for balance, he hauls the *kōzo* through the village to the steaming area. On a productive day, he may harvest and haul about twenty bundles. Kurodani, 1972.

35–37. Steaming. A tightly bound bundle of *mitsumata* shoots is placed upright into a sunken cauldron, from which steam already rises. A tubular barrel is lowered over the shoots, and the mouth edge sealed with mud. The fire is stoked as the steaming progresses. Plate 37 shows the lever system used for lowering and raising the barrel.

38. Two or three people work together to strip the steamed and loosened bark from a number of stalks at once. Stripping complex, triple-forked stalks of *mitsumata* for clean, unbroken bark yield involves a technique somewhat different from that used to strip straight *kōzo* stalks.

Photographs 35–38, obtained from the Ōji Paper Museum, are credited to master papermaker Eishirō Abe and were ostensibly taken at his workshop in Yakumo village, Shimane Prefecture. The date is probably in the 1950s.

39–44. Farmers in Yamanashi Prefecture steam *mitsumata* in a small metal shed (40). Harvested *mitsumata* stalks are kept fresh by standing them in a stream (39). White balls of *mitsumata* flowers shower the ground when the steam shed is opened and the stalks are removed (41). The bark peeling is done by women and older children, and the bare stalks are bundled for use as winter fuel. Bark peeling is usually a lighthearted interlude in the work cycle of both farmer and papermaker; here the mood was amplified by the presence of a cameraman (42, 43).

The peeled bark is dried on racks for several days (44). Yamanashi Prefecture, 1972.

45. Women stripping *kōzo*. Fukuoka Prefecture (papermaking town of Yame?), 1950s.

46–52. This series of *kōzo* steaming and stripping photographs, obtained from the Ōji Paper Museum, unfortunately bore no date and was otherwise unidentified; it was possibly taken in the early 1960s, or before, judging from the setting and the clothes.

The wad of straw rope sealing the edge of the wooden barrel where it meets the metal cauldron lip can be clearly seen in plates 46–48. After removing the barrel (48), the hot *kōzo* stalks are doused with cold water (49).

Making an incision with her thumbnail through the bark at the base of stalk, the grandmother twists and wrenches the skin away from its woody core (50–52). These stalks are just the right length to enable her to do this in one or perhaps two motions.

53. Black bark drying. The overhanging eaves of a large storehouse protect hanging racks of drying bark. Nagano Prefecture, 1972.

#### Soaking Black Bark

54–59. Kudani women use bamboo poles laid across the village stream for support as they trample knotted bundles of black bark (54, 55). The bark has already been soaked in the river for a day, weighted with stones, to soften and loosen the thin, dark, outer skin. Trampling, the next process, not only loosens it further, but also helps soften the thick, white, inner bark. A few generations ago, this work was done with bare feet (54–58). Trampled and partially cleaned bark lies across a wooden bridge (59). Workers will take this moist and soft material to be scraped. Kurodani, 1971.

### Scraping Black Bark

60. Family members gather around a barrel of black bark for *kawa-nuki*, bark scraping. Note the small boards, legged on one side, against which the bark is scraped; the scraper blade set in a piece of wood (foreground of photo) is a variation of a tool used to cut straw. Location unknown; probably 1950s.

61–63. Three variations of bark scraping in Yoshino, 1947. The old and the young woman (61, 63) scrape sideways, at right angles to their legs, while the man sits comfortably crosslegged.

64–66. Long, clean strokes of the blade in the direction of the grain are necessary to scrape bark; this skill is clearly pictured in this series from Yoshino, 1974. The woman uses the same posture as women from the same village three decades ago, shown in plates 61 and 63. The black chaff that piles up around the scraper is often used to make *chiri-gami*. Note the knife and the scraping surface.

67. A shaped bundle of rice straw provides a flexible yet sturdy surface for bark scraping. The worker sits on thick, braided straw extensions of the bundle, which serve to anchor it. A comparison of various types of blades used for scraping as well as the surfaces against which the bark is scraped makes an interesting study in local techniques. Kamikawasaki, Fukushima Prefecture, 1971.

68. This scraper finds that, for some reason, the bark is heavily scarred across the grain. Generally, cuts across the grain shorten and weaken paper fiber. Kurodani, 1971.

69, 70. For fine white papers, like the tissue-thin *Yoshino-gami*, the bark must be cleaned of as much green cuticle as possible. Sitting in a well-lighted place outside, this papermaker gives the white bark detailed inspection. Yoshino, 1972.

71. The amount of time and effort expended in scraping the bark is the individual papermaker's choice. Certain impurities give the paper color and character. The fewer the impurities, the whiter the paper will be. Kurodani, 1972.

72, 73. Care and patience are demanded in every stage of papermaking, but the ability of the older craftsmen and women to relate directly and unconsciously to their materials is not easily emulated by young people today. Alongside the more obvious aspects of economic growth, the concept of work has changed. Min Eguchi of Oguni, Niigata Prefecture, directs close attention to every piece of bark she scrapes. 1973.

74, 75. Women at Tadao Endō's workshop employ mechanical bark scrapers of his invention. They feed the bark strip by strip through rollers lined with sharp blades. Small electric motors controlled by foot pedals power the scraping machines. Shiroishi, Miyagi Prefecture, 1972.

76. A new bark scraper resembles an office machine in design. Gokayama, Toyama Prefecture, 1975.

### Boiling

77, 78. The cauldrons used to boil bark at Yoshino in 1947 did not differ from the ones used for the preceding two or three centuries, and perhaps longer. An inspection of the various tools and implements in plate 77 is a lesson in the use of available materials, though not all their functions are immediately clear.

79. A worker adds bark to a cauldron of steaming water and alkali solution. Location unknown, probably 1960s.

80. At this outdoor cauldron, women remove some boiled bark with sticks to check its progress. When the noncellulosic elements have been more or less digested in the alkali solution and the bark can be crushed between the fingers, the cooking is complete. Location unknown, probably 1950s.

81, 82. One of the large cauldrons for boiling used at Kurodani. 1971.

83. Chisako Horii tends the fire under a cauldron of boiling bark. She fuels it with stripped *kōzo* stalks. Kurodani, 1971.

WASHING AND BLEACHING

84–87. A papermaker arranges bundles of boiled bark in a weir for "river bleaching" (*kawa-zarashi*). The bark soaks in the gently flowing water for about two days in winter. The shape of the spread bundle is referred to as a "chrysanthemum." In plates 86 and 87, the papermaker thrashes the fiber through the river water to rid it of clinging residue. Yoshino, 1974.

88. The clean bark from the river is hung in loose bundles to dry in the sun. This process also bleaches the fiber somewhat. Yoshino, 1974.

89. Washing bark in the village stream. Kurodani, 1971.

90. On the hillside above the valley stream, split bamboo ducts leading from a nearby brook feed a rinsing vat for *kōzo* bark. Kurodani, 1971.

91, 92. For centuries the hills of Mino have witnessed the making of high-quality paper. Here, Kōzō Furuta, on a warm rainy day, checks the bark laid out to wash and bleach in a weir. Warabi, Gifu Prefecture (Mino), 1971.

93–95. Bundles of white bark float for several days in a vat of clear running water at Ichibei Iwano's workshop. The winter cold tightens the fibers and produces the best paper. After shaking a bundle free of water, Iwano checks the quality of the bark. Ōtaki, Fukui Prefecture (Echizen), 1972.

96–100. This compelling series of photos of an old woman carting, hanging, and inspecting washed bark on a rainy winter day epitomizes the reason why few young people today are drawn to this arduous craft. Kurodani, 1971.

PICKING OUT THE DROSS

101–108. Picking out the dross (*chiri-tori*) from the boiled and washed bark is probably the most demanding, mechanical, and dreary of the many steps of papermaking. Warabi, Gifu Prefecture (Mino), 1971.

109. Stirring boiled bark in preparation for *chiri-tori*. Note the box seat and the cloth-lined openwork basket. Running water is essential to this work; in fact, *chiri-tori* is done largely in water. Kurodani, 1971.

110. A pointed or bladed instrument is often used to remove scars, rough areas, and bits of black bark. Hida-Kawai, Gifu Prefecture, 1975.

111. One hand protected by a rubber glove, this woman carefully removes bits of black bark from the white. Her basket is quite deep, holding an afternoon's work. Kurodani, 1971.

112. *Chiri-tori* is slow and painstaking, and the quality of a paper reflects the effort of the women who do this labor. A single worker's output is not high; to produce a large quantity of paper, many workers are necessary. Such a scene as pictured here is rarely observed in paper villages today. Nagano Prefecture, 1965.

113. Women bend over their *chiri-tori* work as if doing laundry. Kurodani, 1971.

114. A faucet is the focus of a quartet of *chiri-tori* workers. (The same workshop, after remodeling in 1976, is pictured in plate 119.) Ōmi-Kiryū, Shiga Prefecture, 1972.

115, 116. A thatched windbreak and lean-to provides some protection for workers doing *chiri-tori* along a riverbank. Yoshino, 1947.

117. In winter, a straw lean-to was often the only protection for *chiri-tori* workers. The boxes serve both as windbreaks to protect the hands and as containers for bark. Seemingly brighter than the drift of snow in the foreground, hanks of cleaned bark sit on shelves. Gifu Prefecture, 1959.

118. A lonely *chiri-tori* hut along a river in northern Honshu, just large enough to accommodate a solitary worker. Tōno, Iwate Prefecture, 1950s.

119. Chika Naruko's workshop, famous for its *gampi* paper, was remodeled in 1976. Ōmi-Kiryū, Shiga Prefecture, 1976.

120. It is largely the efforts of older women doing the mechanical, routine jobs that has allowed Japan to continue producing the greatest quantity and highest quality of handmade paper. Unlike such crafts as ceramics, papermaking is attracting few young people, either in Japan or from abroad. Karasuyama, Tochigi Prefecture, 1972.

BEATING

121, 122. Ichibei Iwano's son beats bark into a pulpy mass. Using a hardwood bat, he beats from left to right and then back again across the wooden slab, occasionally tossing some water on the fiber and turning the mass over. Hand-beating like this is the only kind done at Iwano's workshop—a rarity in the contemporary paper world. Ōtaki, Fukui Prefecture (Echizen), 1972.

123, 124. Kōzō Furuta of Mino uses carved mallets to beat fiber on a granite slab. This "chrysanthemum" pattern is part of the Mino tradition; the carving is not decorative, but increases the surface area of the mallet heads. Warabi, Gifu Prefecture (Mino), 1972.

125. Women of Shiroishi beat fiber pulp with small, carved mallets. Tadao Endō's wife is from Gifu, thus this adaptation of the carved Mino beaters in this workshop of northern Japan. Shiroishi, Miyagi Prefecture, 1950.

126. A couple beats fiber across a hardwood slab supported on a low table. The traditional and exhausting method of hand-beating has nearly disappeared in Japan, which is one reason for its emphasis in these photographs. The rhythm of the beaters inspired numerous papermaking songs, which helped reduce the monotony and strain of the work. Akibatake, Gumma Prefecture, 1971.

127. Two light bats were and still are used in Yoshino to beat fiber pulp. Yoshino, Nara Prefecture, 1947.

128. A waterwheel is still in operation to mill pulp in a remote area of Tosa. Unfortunately, no

photograph was available of the beating mechanisms, but they are described as being simple, cylindrical wooden hammers. Iwahara, Kōchi Prefecture (Tosa), 1973.

129. Probably the best beating technique after hand-beating is that done with wooden hammers driven off an electrically powered camshaft. The hammers rotate as they rise and fall. Kurodani, 1975.

130, 131. Two variations of wooden hammer beaters. Sakai, Nagano Prefecture, 1973.

132. An ingenious mechanical beater, which is distinctive as an object in itself, is now used perhaps once a year to make a special kind of paper. Kaminoyama, Yamagata Prefecture, 1973.

133. A papermaker carries freshly cleaned bark fiber to the beater. Behind her is a large Hollander-type beater. There are two types of Hollander: one has a rotating drum with metal teeth, like a cogwheel; the other is fitted with scythelike, curved blades (see p. 96). The pulp solution flows down from the Hollander into the adjacent trough. Kurodani, 1971.

134. A worker removes pulp encrusting the area around a Hollander driveshaft. The pulp solution —fiber and water—circulates in the oval tub as it is fed through the rotating teeth or blades of the Hollander. Kurodani, 1971.

135. The metal sluice on the upper left leads from a Hollander. The pulp in the draining vat must be stirred with a paddle to loosen fibers from the screen below and allow the water to run off. Kurodani, 1971.

136. While it is draining, the pulp is inspected for dirt and dross. This is the final inspection point. A papermaker gathers up the last of the pulp from a vat. Kurodani, 1971.

137, 138. Ichibei Iwano gives the beaten pulp a final rinsing in ice-cold water, churning the pulp in a cloth supported by upright bamboo poles. This process further whitens and cleans the fiber. Ōtaki, Fukui Prefecture (Echizen), 1972.

Mucilage Preparation

139. The amazing ingredient in Japanese papermaking is the mucilage, called variously *neri*, *nori*, *nire*, and the like. Here cleaned *tororo-aoi* roots are being pounded with wooden mallets to obtain the slippery mucilage. Misumi, Shimane Prefecture, 1972.

140. Pounding *tororo* roots, Kurodani, 1959.

141. The setting indicates that this is probably a farmer-papermaker's workshop. The barefooted woman is smashing *tororo* roots in a massive stone mortar. Only part of the mucilage is released when the root is smashed; most will flow out as the roots rest in buckets of cold water. Location unknown, probably 1950s.

142. A papermaker lifts mucilage-producing vegetable matter from an underground storage vat. The plant material will keep for about one year this way, floating in a phenolic preservative solution. Akibatake, Gumma Prefecture, 1971.

143. Peeling the bark of the *nori-utsugi* tree to make mucilage. *Nori-utsugi* produces a better mucilage than *tororo*, but is not as widely used for many reasons. It comes almost entirely from Hokkaido. Yoshino, 1947.

144. A modern *tororo* crusher. Kurashiki, Okayama Prefecture, 1973.

145. The millstone principle used for a *tororo* crusher. The stone crushers rotate, and the mucilage escapes through the metal screen and is drained off. This machine is no longer in use. Kaminoyama, Yamagata Prefecture, 1973.

146–148. Processing mucilage before adding it to the pulp vat. The *neri* is first squeezed through a bag, and the viscous liquid is caught in a bucket. Then, the contents of the bucket are poured through two filter bags, one inside another (148). Ino, Kōchi Prefecture (Tosa), 1973.

Scooping and Couching

149. Mucilage may be refined (preceding plates) and ladled into the scooping vat solution or it may be squeezed through a filter bag directly into the vat, as pictured here. In fact, this step comes after pulp has been added to the vat and the pulp solution stirred and agitated (plates 150–153), but the photograph is included at this point to give some visual continuity to the various sequences. The solution is stirred again after the *neri* has been added. Yoshino, Nara Prefecture, 1947.

150. Kōzō Furuta's wife stirs the pulp solution with a pole to disperse the fibers evenly preparatory to scooping. The pulp to water ratio is about one kilogram of dry fiber to 100 kilograms (100 liters) of water, but this varies according to the type of paper being made. The viscosity of the solution is evident in the way the liquid sloshes when stirred. The hooked cord supporting the stirring pole is fastened to a flexible pole on the ceiling; such cords are used primarily to carry some of the weight of the scooping mold when it is filled with solution. Warabi, Gifu Prefecture (Mino), 1972.

151–153. After the initial stirring with a pole, a comblike agitator is lowered into the vat. The agitator is pulled and pushed vigorously through the solution to further refine it and disperse the fibers. Kurodani, 1972.

154–158. The mold (*sugeta*) has two parts: a hinged wooden frame (*keta*) and a finely woven bamboo screen (*su*) held in place by the frame. The papermaker scoops solution into the mold, carefully balancing its heavy weight. The cords fastened to the ceiling, which help support the mold, are clearly visible here. Quickly and rhythmically, pulp solution is tossed back and forth across the *su*. The pulp solution rolls across the screen "like the waves of the sea," the papermakers say. When the desired thickness is attained, the worker briskly discharges excess solution over the back edge of the mold. The mold is then rested on two slats over the vat, and, as water drains through the screen, the papermaker carefully picks unwanted matter out of the newly formed sheet. The mold is then opened to remove the screen with the wet sheet. Kurodani, 1972.

159–166. The weight of a large screen, such as the one pictured, filled with pulp solution is formidable. Even with four supporting cords, the strain of handling such a weight is great and demands enormous expertise to handle throughout a working day. Further, if the mold is not carefully and properly balanced, the papermaker is put under additional strain to keep the pulp solution uniform and to allow an even deposition of fibers on the screen. Though the paper being made in this series of photographs is not highly refined—a type of *unryū-shi* with rough fibers—the papermaker concentrates intensely and is totally absorbed in his work. Karasuyama, Tochigi Prefecture, 1972.

167, 168. Yoshino workshops in 1947 and 1973. The seated scooping technique seen just thirty years ago has been almost completely abandoned in Japan today. Naturally, seated work can only be done with a small mold and vat.

169, 170. These two studies of scooping in Yoshino in 1947 deserve close scrutiny. Isolation, the natural conservatism of craftsmen, and, until recently, poverty preserved the craft in Yoshino as it was practiced for several hundred years. Even today, innovations are few. Though the equipment appears rough and primitive, there is nothing slipshod in the work or the product—the techniques are precise and efficient, yet labor-intensive. The paper made is of the best quality. Yoshino is famous for the production of highly refined, white tissues, whose strength, in spite of their gossamer thinness, has always amazed Westerners (plates 199, 200).

171–176. A huge screen for *fusuma* paper is used at the workshop of the late Heizaburō Iwano (no relation to Ichibei Iwano). Instead of dipping the mold into the vat solution, a bucket is used to pour the solution onto the screen. The great weight of the mold and solution is evident from the thickness of the supporting cords or wires, which balance the mold in the middle and allow the screen to be tipped back and forth. Since most of the weight is supported from above, one woman takes a moment to relax as the other worker lifts her end of the mold (174). The solution is allowed to roll from one end to the other repeatedly, with some shaking of the screen by the workers. Some excess solution may be released from either end of the mold, but largely the liquid is allowed to drain through the screen. The scooping is rather slow and ponderous; the quality of such big sheets is usually not as high as the paper made by a single papermaker controlling a mold by himself. This workshop has an even larger screen, which is handled by four people, but it is not often used. Ōtaki, Fukui Prefecture (Echizen), 1972.

177. This mold is constructed so that twenty postcards are scooped together. When the screen is released from the mold, the postcards will be deposited onto twenty neat piles. Kurodani, 1971.

178–180. The rhythm and movements of scooping are dictated by the kind of paper being made, the traditions of the various regions, and, of course, by the dispositions of the individual papermakers. Some indication of this variety of rhythm and movement is seen in these three photographs showing what appears to be excess solution being discarded at the end of the scooping cycle but in fact is the wave of solution cresting on the edge of the mold before sloshing back across the screen. Note that the molds in all three cases are tilted down toward the papermaker. Yoshino, Nara Prefecture, 1974; Ino, Kōchi Prefecture (Tosa), 1976; Warabi, Gifu Prefecture (Mino), 1972.

181. Kōzō Furuta eliminates impurities from a sheet of *Mino-gami*. Warabi, Gifu Prefecture (Mino), 1972.

182. Removing air bubbles from *Mikawa Morishita* paper by gently drawing the air up with a straw. Obara, Aichi Prefecture, 1975.

183–185. When the securing frame of the mold is raised, the screen on which the newly laid paper rests can be lifted out of the mold. The screen ribs and flecks of fiber in the *unryū-shi* are silhouetted against the sunlight from the windows (184, 185). Karasuyama, Tochigi Prefecture, 1972.

186–191. A papermaker prepares to transfer the newly formed sheet from the screen to the growing stack of fresh, wet sheets. The separation in the middle of the large screen allows two sheets to be made at one scooping. The screen edge is carefully aligned with the stack of paper, and the screen is slowly lowered upon it. As her right hand lowers the screen onto the stack, her left hand holds the edge that touched the stack first firmly in place (190). When the screen lies flat on the stack, she uses her right hand to grasp the edge closest to her, wiggles the edge to free it from the paper, and then swiftly draws the screen up and away from the wet paper. Ogawa, Saitama Prefecture, 1972.

192–194. Accumulation of experience intensifies the craftsman's concentration and commitment to work as well as enhances the ease with which work can be done. Casualness and strain are, on the other hand, the prerogatives of the young. Clearly displayed in these photographs is the intensity of involvement in the relatively simple task of depositing newly laid paper on the screen onto the stack of wet sheets; the woman pictured probably has close to a half-century of experience in paper-making. Ōtaki, Fukui Prefecture (Echizen), 1972.

195–198. A block and tackle or winch system is needed to lift the giant screen for *fusuma* paper, and two workers must help in stacking the new sheets. Ōtaki, Fukui Prefecture (Echizen), 1973.

199–201. Tissue-thin *Yoshino-gami* sheets are not stacked, but are couched directly on the drying boards. The square and rectangular patches on the boards cover cracks in the wood. Because boards swell with the moisture from the paper and contract again as it dries, any cracks that have formed in the boards open up as the wood contracts and will split the paper. To avoid this, patches of paper are pasted over such cracks, providing an unblemished surface for the paper sheets. Yoshino, Nara Prefecture, 1947 (plates 199, 200) and 1973 (plate 201).

PRESSING

202, 203. The simplest type of paper press; stones are added as the water is pressed out. Yoshino, 1947; Takaoka, Kōchi (Tosa), 1973.

204–206. The next step in sophistication—stone-weighted lever presses. The stone pile on the end of the lever (204) becomes increasingly hard to balance as stones are added to increase pressure. The answer to this problem is the somewhat more sophisticated lever press in plate 205. Some stone weights for a lever press are shown in plate 206. Yoshino, Nara Prefecture, 1947; Ōmi-Kiryū, Shiga Prefecture, 1972; Ōtaki, Fukui Prefecture (Echizen), 1973.

207. A modern version of the lever press employs a large drum and a steel "I" beam. The drum is partially filled with water at first; more water is gradually added to increase the pressure. Takaoka, Kōchi Prefecture (Tosa), 1973.

208. Mechanical presses of various kinds are now the most widely used. Many papermakers are quite content with a homemade screw press; though there are many presses of this type in Japan, probably no two are identical. Yoshino, Nara Prefecture, 1972.

209, 210. Presses of various degrees of technical refinement are employed; these two photographs illustrate a screw jack press and a powerful hydraulic jack press. Regardless of whether the pressing is done by a pile of stones or by the most modern device, the application of even and gradual pressure is important—the water must be pressed out evenly and gently, and the center of the stack of sheets must not retain more moisture than the edges. Karasuyama, Tochigi Prefecture, 1972; Ogawa, Saitama Prefecture, 1972.

DRYING

211, 212. Holding a camellia leaf in his mouth, a papermaker brushes sheets of paper onto a drying board. He uses the leaf to smooth out the corners of the thick sheets on the boards (see p. 104).

213. A papermaking couple carry boards covered with damp sheets into the sun. Sunlight and wind dry the paper quickly and make it crisp. Kurodani, Kyoto Prefecture, 1971.

214, 215. Two more scenes in Kurodani. The tall pole racks in the background of plate 214 are to dry rice straw or other farm produce; single-pole racks, such as seen in the foreground, are props for drying boards. Drying board storage sheds dot the hillsides above the town. Kurodani, Kyoto Prefecture, 1971.

216, 217. Brushing paper onto drying boards, Yoshino, 1947 and 1974. The brush in the older picture is made of palm fibers clamped into a split stalk of bamboo. A soft paint brush is used today.

218–220. From years of use, the grain of the drying boards becomes etched in low relief. Such etched boards leave grain impressions on the dried sheets. The front rows of boards in plate 219 are laid at a low angle to catch a maximum amount of winter sun.

The long, narrow sheets on the boards are probably *gasenshi*. When the sheets are dry, they are easily peeled from the boards. The cloth on top of the growing stack of sheets being taken from the boards weights down the curling paper (220). Both sides of the boards are covered with new paper; when one side dries and the paper is removed, the boards are turned over to allow the paper on the reverse side of the boards to dry. Yoshino, Nara Prefecture, 1974.

221, 223. For very large, soft sheets of *fusuma* paper (see plates 171–176), Heizaburō Iwano devised a cabinet dryer. The drying boards slide into channels, and the entire cabinet is heated. The large sheets dry as quickly in the cabinet as they do in the sun. Ōtaki, Fukui Prefecture (Echizen), 1973.

223, 224. The common iron plate dryer is shaped like a narrow, sharply peaked tent. The central cavity contains the heat source; in this case, the dryer is wood burning. Wet paper sheets are brushed onto the metal surfaces of the two long sides; steam starts rising from them immediately, and they dry very quickly. Such drying, it is thought, frustrates natural shrinkage and leaves the paper somewhat flaccid, but on rainy days it is eminently practical. There are other variations of the metal plate dryer (see p. 103). Kurodani, Kyoto Prefecture, 1971.

225. Trimming off the deckle edges. Note the special knives and their handles, flattened on one side. To cut a true edge through a stack of paper takes no small skill, preceded by great patience with whetstone and knife. Yoshino, Nara Prefecture, 1947.

226. Burying a stack of paper in the snow to bleach. The woven straw "coat" the papermaker is using is still worn. Oguni, Niigata Prefecture, 1975.

227. A Yoshino papermaker carries finished paper to market. As late as 1947, transportation and cartage of the seventeenth and eighteenth centuries—carrying pole and river ferry—were still alive there. From the moment the *kōzo* was cut to the arrival of the finished paper at market or its delivery to a wholesaler, each sheet received the full measure of the papermaker's effort, energy, and commitment.

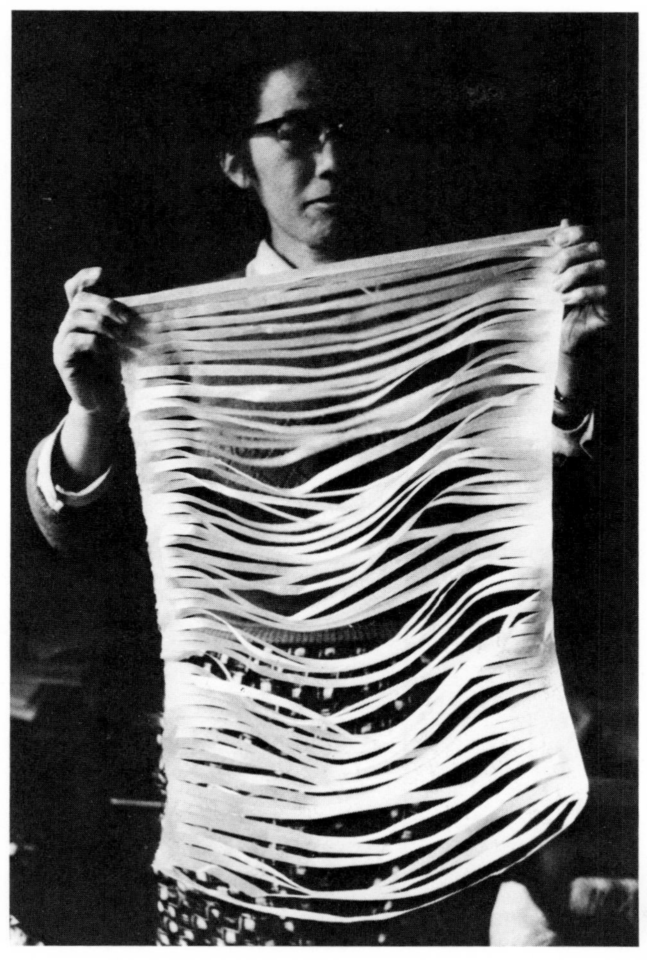

228–233. Michiko Kawaguchi, a weaver in the northern Tamba area near Kurodani, has researched and revived *shifu*—woven paper—techniques. Plate 228 shows a paper sheet cut to make *shifu* "yarn." The margin on the long sides of the sheet will be torn on alternate sides every two strips in order to form one long strip (229). This strip is spun on a spinning wheel to form yarn (230–232), which is then woven into textiles (233). The finest *shifu* is woven with a silk warp, but a cotton warp is more common.

229

230

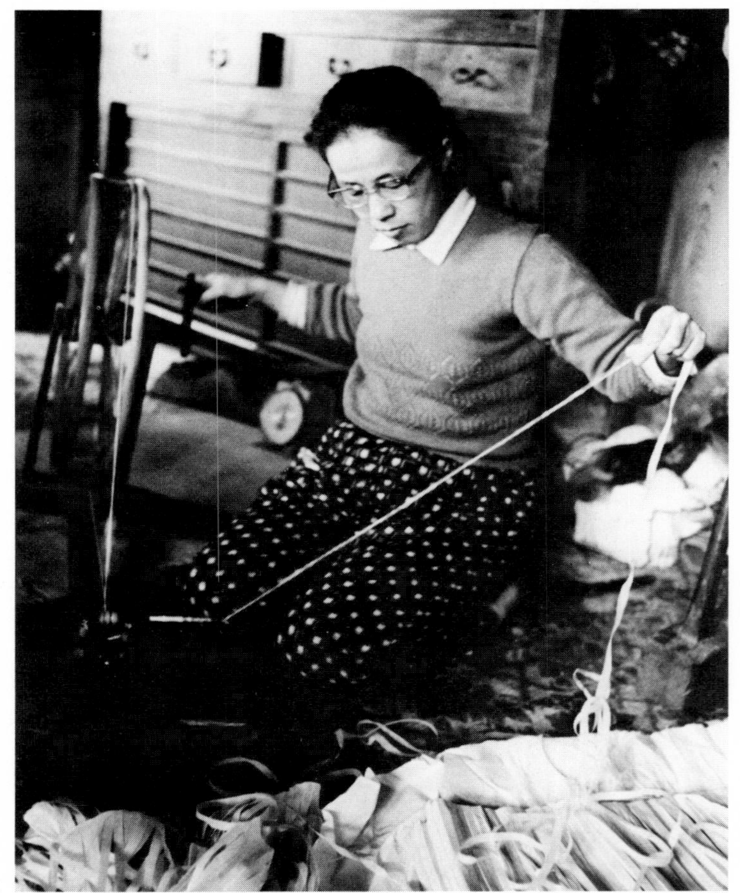

231

232. *Shifu* paper strip and yarn.

233. *Shifu* kimono and *obi* (right) fabrics.

234

235

234–236. The making of *su*, the screens used in papermaking, is a highly exacting art. In Warabi, Gifu Prefecture, a craftsman weaves a *su* out of thinly split and beveled bamboo splints. Bobbins carry the warp threads; this technique is basically the same as that used to weave the reed facing of tatami mats, though much more delicate. In order to form single, continuous strands of bamboo extending the length of the screen, the beveled ends of short bamboo splints are butted together. Papermakers fear that the technique of making *su* will be lost, since the older craftsmen are ceasing to work and no young people are drawn to this craft.

# Old Provinces

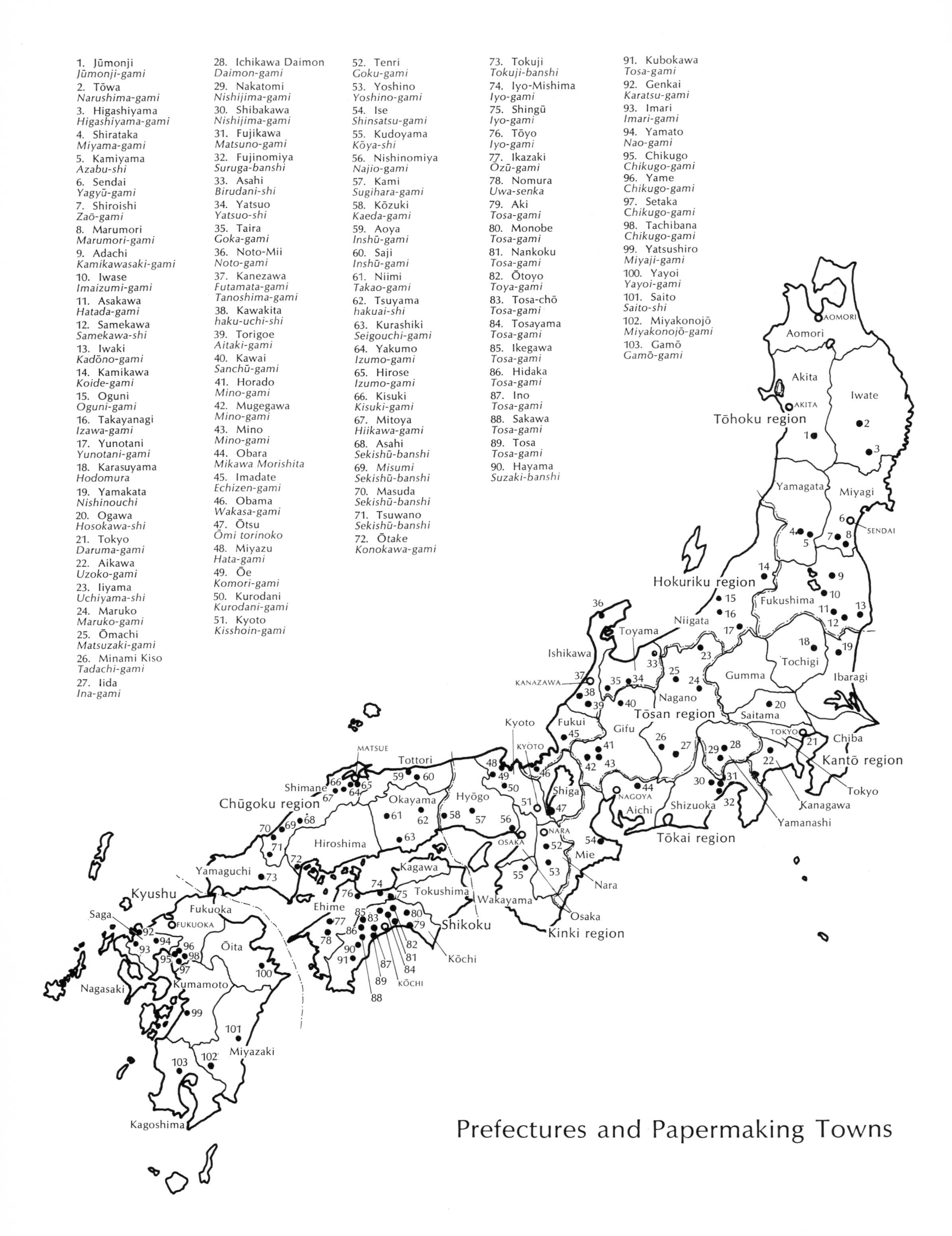


Prefectures and Papermaking Towns

# Glossary

## English

CHAIN LINES. Impressions left on the paper by the structure of the screen; a natural watermarking of the screen's structure in the paper.

COUCH. To transfer a freshly made sheet of paper from the screen-mold to a flat surface or to a pile of wet sheets; this pile is sometimes called the couch (also post).

DECKLE. Detachable or loose wooden frame that holds the papermaking screen; the edges of the deckle contain and confine the pulp. In Japan the deckle forms the upper half of the mold.

DECKLE EDGE. Rough edges of a sheet, formed naturally as pulp washes against the deckle or mold top.

FILLER. *See* LOAD.

FURNISH. Paper material "furnished" to the beater; paper pulp.

HALF-STUFF. Pulp washed and somewhat bruised, ready for the beater; this term is generally used for rag papers and has essentially the same meaning as furnish.

HOG. Agitator, usually mechanical and having fanlike blades, which churns the pulp solution and keeps the fibers evenly distributed.

HOLLANDER. Mechanical beater of Dutch origin; bast solution circulates around an oval tub and through an iron drum fixed with blades.

LAID PAPER. Paper made with the chain lines showing.

LOAD. Clays or minerals added to the pulp solution to slicken the paper's surface and/or to add weight to the finished sheets.

POST. Stack of newly made, wet sheets. In Western papermaking, the sheets are separated by felts.

RETTING. Exposing vegetable fiber to moisture in a process of partial decay in order to loosen fiber from its woody core and begin the breakdown of noncellulosic constituents.

"RICE PAPER." Misnomer for Oriental and especially Japanese handmade papers. The original "rice paper" is a paperlike substance sliced from the pith of the Taiwanese *Aralia papyrifera* tree.

SIZE OR SIZING. Gelatinous or glutinous preparation made from glue, starch, alum, etc. and used for filling the pores of paper. The most common use of sizing is to decrease the absorbency of papers to be used with ink or pigments.

STOCK. Pulp solution ready to be formed into sheets of paper; or, wet pulp in any of various stages of processing.

"VATMAN'S STROKE." As used by Dard Hunter, this term refers to the various methods used to toss the pulp solution in the mold.

WATERLEAF. Unsized paper.

WATERMARK. Design applied to the surface of the paper screen, which reproduces in the paper as translucent areas thinner than the paper sheet. In Japan, such a design may also be applied to silk gauze (see *sha*) laid over the screen.

### JAPANESE

CHIRI. Bits of discarded matter; dross; in papermaking, pieces of black bark and injured fiber.

CHIRI-TORI. Removing *chiri* from bark that has been boiled and washed.

DŌSA. Sizing solution made from animal glue and alum.

FUNE (also *sukibune*). Papermaking vat. Literally, "boat" or "papermaking boat."

GAMPI. Bast fiber shrub of the *Thymelaeaceae* family, characterized by tough, long, fine, glossy fibers.

HANKUSA. "Half-grass," that is, half fiber and half pulp paper.

JŌ. Unit of measure for paper; for *hanshi*, one *jō* equals 20 sheets; for *Mino-gami*, 48 sheets, etc.

KAJINOKI. Paper mulberry; *Broussonetia papyrifera*, Vent.

KAMI. Paper; as a suffix, often pronounced *-gami;* the same character is also read *-shi*. Custom determines which pronunciation is used.

KAMISUKI. Papermaking.

KAN. Traditional Japanese weight unit; one *kan* is 3.75 kilograms.

KAWA-MUKI. Removing the dark outer skin of bark with a scraper or knife.

KAWA-ZARASHI. "River bleaching"; washing and soaking boiled bark in the gentle current of a river to clean and bleach it.

KETA. Hinged, wooden frame that holds the screen (*su*); used in scooping paper; the upper part may have handles and is comparable to the Western deckle.

KIN. Traditional Japanese weight unit; one *kin* equals 600 grams.

KIZUKI (or *shōkusa*). Pure-fibered paper; unadulterated.

KOBAN. Small-sized sheet of paper.

KŌZO. General term for a variety of papermaking mulberries, characterized by strong, sinewy, and long fibers.

KUROKAWA. "Black bark"; bark without the outer, dark skin removed.

MARU. Unit of measure for paper; for *hanshi*, one *maru* equals 6 *shime* (q.v.), or 12,000 sheets; for *Mino-gami*, 4 *shime*; for *hōsho*, 10 *soku* (q.v.); for *Sugihara-shi*, 8 *soku*, etc.

MASE (or *zaburi*). Agitator, shaped like a huge comb, for stirring the pulp solution in the vat; hand operated.

MIMITSUKI. Literally, "with ears attached"; paper with untrimmed deckle edges.

MITSUMATA. Bast fiber shrub of the *Thymelaeaceae* family, characterized by fine-grained, soft, absorbent, and slightly lustrous fibers.

NAGASHI-ZUKI. "Discharge papermaking"; method of papermaking peculiar to Japan, characterized by ejecting excess pulp solution from the mold; made possible by the use of *neri*.

NERI. General term for the various kinds of vegetable mucilage used in Japanese papermaking, such as that from the *tororo-aoi* root.

NORI-UTSUGI. Tree of the saxifrage family; its inner bark is used to make one type of *neri* mucilage.

ŌBAN. Large-sized sheet of paper.

OMOTE. Front or smooth surface of a sheet; the side next to the screen and to the drying surface.

SARASHIKO. Bleaching powder; chloride of lime.

SHA. Fine screen, usually silk gauze; a *sha* is placed over the screen in papermaking either to create "wove" papers or to hold a watermark.

SHIBU. Astringent juice of unripe persimmons; persimmon tannin; used to waterproof and strengthen paper.

SHIME. Unit of measure for paper; for *hanshi*, one *shime* equals 10 *soku* (q.v.), 100 *jō*, or 2,000 sheets.

SHIROKAWA. White, inner bark from which most washi is made.

SHITO. "Post" (q.v.).

SHŌKUSA. *See* KIZUKI.

SHŌSŌ-IN. Imperial Repository in Nara (built in 752), which houses treasures of Emperor Shōmu (reigned 724–749) and other material of the Nara period (710–794).

SHUKUSHI (also *sukushi*, *sukigaeshi*). Recycled paper.

SOKU (also *taba*). Unit of measure for paper; for *hanshi*, equals 10 *jō*, or 200 sheets.

SU. Flexible screen, usually bamboo but sometimes reed, that acts as the sieve or strainer upon which paper is formed.

SUGETA Papermaking mold; combination of *keta* (frame) and *su* (screen).

SUKASHI Watermark.

SUKIBA. Papermaking work area or workshop.

SUKIBUNE. *See* FUNE.

SUKIGAESHI. *See* SHUKUSHI.

SUKU. To make paper; to (spread out) thin.

SUKURYŪSHIKI. Mechanical "hog" agitator.

SUKUSHI. *See* SHUKUSHI.

SUTEMIZU. Cast-off excess pulp solution; the characteristic of the *nagashi-zuki* technique.

TABA. *See* SOKU.

TAME-NAGASHI (-ZUKI). Papermaking method combining *tame-zuki* and *nagashi-zuki* techniques.

TAME-ZUKI. "Accumulation papermaking." The ancient, original technique of scooping—now considered Western by Japanese scholars—in which sheets are formed very quickly and without the use of mucilage by allowing the pulp solution to drain through the screen.

TENPI-ZARASHI. "Sun bleaching" of bark.

TESUKI WASHI. Hand-laid (-made) paper.

TORORO-AOI. Plant of the mallow family; its roots are crushed to make the most common kind of *neri* mucilage.

TSURI (also *yumi*). Flexible bamboo poles fastened to the ceiling above the vat; cords attached to the ends of the *tsuri* help support the mold.

UCHIKATA. Method of hand-beating paper fiber.

URA. Back or bottom surface of a sheet; the upper side in the mold; this is also the side that receives the brush when the sheet is applied to a flat drying surface.

YŌSHI. Western machine-made paper.

YUKI-ZARASHI. "Snow-bleaching" of bark or paper.

YUMI. *See* TSURI.

ZABURI. *See* MASE.

# Bibliography

American Paper & Pulp Association, ed. *The Dictionary of Paper*. 2nd ed. New York: American Paper and Pulp Association, 1951.

Barrow, William J. "Physical and Chemical Properties of Book Papers, 1507–1949." *Permanence/Durability of the Book*. Richmond, Virginia: W.J. Barrow Research Laboratory, Inc., 1974.

Boger, H. Batterson. *The Traditional Arts of Japan*. Garden City, N.Y.: Doubleday, 1964.

Britt, Kenneth. *Handbook of Pulp and Paper Technology*. New York: Van Nostrand Reinhold, 1970.

Bunkachō, ed. *Tesuki washi: Echizen hōsho, Sekishū-banshi, hon Mino-gami*. Tokyo: Dai-ichi Hōki, 1971.

Burkhart, Timothy. *Kitefolio*. Berkeley: Double Elephant, 1974.

Carey, James P. "Chemistry and Chemical Technology." *Pulp and Paper*. Vol. III. New York: Interscience Publishers, Inc., 1961.

Carter, Thomas Francis. *The Invention of Printing in China and its Spread Westward*. New York: Columbia University Press, 1925.

Casal, V.A. *The Doll Festival*. Paper read at the Kobe Women's Club, 1938.

Clapperton, R.H. *Paper: An Historical Account of Its Making by Hand from the Earliest Times Down to the Present Day*. Oxford: The Shakespeare Head Press, 1934.

Crawford, T.S. *A History of the Umbrella*. New York: Taplinger, 1970.

Elliott, Harrison, collector. The Harrison Elliott Collection of Paperiana, Rare Book Division, Library of Congress. (This section of a room includes frames, utensils, photos, samples, and mounds of unorganized information on paper.)

Engel, Heinrich. *The Japanese House—A Tradition for Contemporary Architecture*. Tokyo and Rutland, Vt.: Tuttle, 1964.

Fedderson, Martin. *Japanese Decorative Art*. New York: Thomas Yoseloff, 1962.

Fobel, Jim and Boleach, Jim. *The Stencil Book*. New York: Holt, Rinehart & Winston, 1976.

Gotō, Seikichirō. *Nihon no Kami: Japanese Hand-made Paper*; also entitled *Japanese Paper and Paper-making*. Trans. by Iwao Matsuhara. Vol. I, Eastern Japan; Vol. II, Western Japan. Tokyo: Bijutsu Shuppansha, 1958–60.

———. *Washi no furosato*. Tokyo: Bijutsu Shuppansha, 1959.

Haibara, Naojirō. *A Collection of One Hundred Japanese Handmade Papers, including the Most Famous Makers*. Tokyo: 1935.

Hanada, Tokutarō and Narita, Kiyofusa. *Kami-suki uta*. 2 vols. Tokyo: Seishi Kinenkan, 1951.

Hart, Clive. *Kites*. New York: Praeger, 1967.

Hashimoto, Sumiko. *Japanese Accessories.* Tokyo: Japan Travel Bureau.

Herrigel, Eugen. *The Method of Zen.* Ed. by Hermann Tausend. Trans. by R.F.C. Hull. London: Routledge and Kegan Paul, 1960, 1969.

Hisamatsu, Shin'ichi. *Zen and the Fine Arts.* Trans. by Gishin Tokiwa. Tokyo: Kodansha International, 1971.

Huang, Al Chuang-liang. *Embrace Tiger, Return to Mountain.* Moab, Utah: Real People Press, 1973.

Hunter, Dard. *A Papermaking Pilgrimage to Japan, Korea, and China.* New York: Pynson Printers, 1936.

———. *Chinese Ceremonial Paper.* Chillicothe, Ohio, 1937.

———. *Old Papermaking in China and Japan.* Chillicothe, Ohio, 1932.

———. *Papermaking, The History and Technique of an Ancient Craft.* New York: Knopf, 1943, 1947.

———. *Papermaking Through Eighteen Centuries.* New York: Wm. Edwin Rudge, 1930.

Ikeda, Toshio, ed. *Izumo no kami.* Iwasaka, Shimane Pref.: Eishirō Abe private pub., 1952.

Joya, Mock. *Things Japanese.* Tokyo: Tokyo News Service, 1958, 1960.

Jugaku, Bunshō. *Paper-Making by Hand in Japan.* Tokyo: Meiji Shobō, 1959.

———. *Nihon no kami.* Tokyo: Yoshikawa Kōbunkan, 1969.

Kaempfer, Engelbert. *The History of Japan.* Trans. by J.G. Scheuchzer. London: Thomas Woodward and Charles Davis, 1728; Glasgow: J. MacLehose and Sons, 1906.

Kannō, Shin'ichi. *Shiroishi-gami* ("Shiroishi Paper"). (In Japanese with some English). Tokyo: Bijutsu Shuppansha, 1966.

Kawakita, Michiaki. *Contemporary Japanese Prints.* Trans. by John Bester. Tokyo: Kodansha International, 1967.

Kennedy, Malcolm D. *A Short History of Japan.* New York: New American Library, 1964.

Kimura, Seichiku. *Shinsen shikan.* Orig. ed., 1777; reprint, Tokyo: Kami no Hakubutsukan, 1977.

Kume, Yasuo. *Tesuki washi senzui* (*The Essence of Japanese Paper: Its History and Habitat*). (In Japanese with some English). Tokyo: Kodansha, Ltd., 1975.

Kung, David. *The Contemporary Artist in Japan.* Honolulu: East-West Center Press, 1966.

Kunisaki, Jihei. *Kamisuki chōhōki—A Handy Guide to Papermaking*, 1798. Translated and adapted by Charles E. Hamilton. Berkeley: Berkeley Book Arts Club, University of California, 1948.

Kuo, Nancy. *Chinese Paper Cut Pictures.* New York: Transatlantic, 1974.

Kurodani Washi Kumiai. *Kurodani no kami.* Maizuru: Kurodani Washi Kumiai, 1967.

Lee, Sherman. *Tea Taste in Japanese Art.* The Asia Society, 1963.

Libby, C. Earl. *Pulp and Paper Science and Technology.* Volumes I and II. New York: McGraw-Hill, 1962.

Mainichi Newspapers Special Editorial Staff. *Tesuki washi taikan* (*A Collection of One Thousand Handmade Japanese Papers*). (In both Japanese and English). Tokyo: Mainichi Newspapers, 1973–74.

———. *Tesuki washi*. Tokyo: Mainichi Newspapers, 1975.

Mason, Joseph W.T. *The Meaning of Shinto*. Kennikat, 1968 (orig. 1935).

Mayer, Ralph. *The Artist's Handbook*. New York: Viking, 1941.

Mikami, Takahiko and McDowell, Jack. *The Art of Japanese Brush Painting*. New York: Crown, 1961.

Munsterberg, Hugo. *The Folk Arts of Japan*. Tokyo and Rutland, Vt.: Tuttle, 1959.

Muraoka, Kageo and Okamura, Kichiemon. *Folk Arts and Crafts of Japan*. Tokyo and New York: Weatherhill/Heibonsha, 1973.

Narita, Kiyofusa. *A Life of Tsai Lung and Japanese Papermaking*. Tokyo: The Dainihon Press, 1966.

Nishikawa, Issōtei. *Floral Art of Japan*. Tokyo: Japan Travel Bureau, 1949.

Noguchi, Isamu. *A Sculptor's World*. New York: Harper & Row, 1968.

Noma, Seiroku. *The Arts of Japan*. Trans. and adapted by John Rosenfield. Tokyo: Kodansha International, 1966.

Oguri, Sutezō. *Nihonshi no hanashi*. Tokyo: Waseda Daigaku, 1953.

Okakura, Kakuzō. *The Ideals of the East*. Tokyo and Rutland, Vt.: Tuttle, 1970.

Ono, S. *Shinto: The Kami Way*. Tokyo and Rutland, Vt.: Tuttle, 1962.

Parkes, Harry S. *Reports on the Manufacture of Paper in Japan*. Presented to the British Parliament, 1871. London.

Pomeroy, Charles A. *Traditional Crafts of Japan*. Tokyo and New York: Weatherhill, 1967.

Rein, J.J. *Industries of Japan*. London: Hodder & Stoughton, 1889.

Reischauer, Edwin O. and Fairbank, John K. *East Asia: The Great Tradition*. Vol. I. Boston: Houghton Mifflin, 1958, 1960.

Reischauer, Edwin O. *Japan, Past and Present*. New York: Knopf, 1964.

Rothenstein, Michael. *Relief Printing*. New York: Watson-Guptill, 1970.

Runes, Dagobert and Schrickel, Harry G., eds. *Encyclopedia of the Arts*. New York: Philosophical Library, 1946.

Shiroishi Laboratory for Local Industrial Arts. "Japanese Paper, *Kamiko* and *Shifu* Fabric" (booklet). Shiroishi, Miyagi Pref., 1964.

———. "*Shifu* Fabric" (booklet). Shiroishi, Miyagi Pref., 1946.

Sindall, R.W. *The Manufacture of Paper*. London: Constable, 1908, 1919.

Stephenson, J. Newell. *Manufacture and Testing of Paper and Board*. New York: McGraw-Hill, 1953.

Stern, Harold P. *Master Prints of Japan*. New York: Abrams, 1969.

Stevens, Richard Tracy. *The Art of Papermaking in Japan* (short monograph). New York: privately printed, 1909.

Streeter, Tal. *The Art of the Japanese Kite*. New York and Tokyo: Weatherhill, 1974.

Tindale, Thomas Keith and Harriett Ramsey. *The Handmade Papers of Japan*. Tokyo and Rutland,

Vt.: Tuttle, Bunshōdō & Benridō, 1952.

Tokyo Doll School. *The World of Japanese Dolls*. Tokyo and Rutland, Vt.: Tuttle, 1962.

Tsien, T.H. *Written on Bamboo and Silk*. Chicago: University of Chicago, 1962.

Tsunoda, Ryusaku; de Bary, Wm. Theodore; and Keene, Donald. *Sources of Japanese Tradition*. New York and London: Columbia University, 1958.

Van Briessen, Fritz. *The Way of the Brush: Painting Techniques of China and Japan*. Tokyo and Rutland, Vt.: Tuttle, 1962.

Von Boehm, Max. *Dolls and Puppets*. New York: Cooper Square, 1966.

Von Hagen, Victor Wolfgang. *The Aztec and Mayan Papermakers*. New York: J.J. Augustin, 1944.

Watanabe, Michitarō. *Washi ruikō*. Tokyo: Diamond-sha, 1933.

Watsuji, Tetsurō. *Climate and Culture, a Philosophical Study*. Trans. by Geoffrey Bownas. Tokyo: Hokuseidō, 1961.

Wheelock, Mary E. *Paper, Its History and Development* (booklet). Chicago: American Library Assoc., 1928.

Wheelwright, William Bond. *Practical Paper Technology*. Cambridge, Mass.: M.J. & W.B. Wheelwright, 1951.

White, Oswald. *Japanese Papermaking*. Diplomatic and Consular Reports. London: Foreign Office, 1905.

Yamada, Sadami and Itō, Kiyotada. *New Dimensions in Paper Craft*. Trans. by Richard L. Gage. Tokyo, New York and Rutland, Vt.: Japan Publications Trading Co.

Yamamo-chō, Gunka-gun, Ibaragi Prefecture Bunka Hozon Kenkyūkai, ed. and pub. *Nishino-uchi-shi*. Ibaragi Pref., 1966.

Yoshida, Tōshi and Rei, Yuki. *Japanese Printmaking: A Handbook of Traditional and Modern Techniques*. Tokyo and Rutland, Vt.: Tuttle, 1966.

Yanagi, Munemichi. "Creative Art from Paper," *This Is Japan*, No. 18.

Yanagi, Sōetsu. *The Unknown Craftsman*. Translated and adapted by Bernard Leach. Tokyo: Kodansha International, 1972.

———. "Washi no bi." *Ginka* 9 (Spring, 1972).

# Index

PHOTO CREDITS

Martha Guthrie, 6, 32–34, 53, 58, 68–70, 81–83, 91–95, 101–08, 120–24, 137, 138, 150, 159–66, 171–76, 180, 181, 183–94, 208–13, 215, 230–38
Sukey Hughes, 54–57, 59, 74, 75, 90, 109, 113, 114, 133, 139, 177, 205, 214, 223, 224
Kokan Tabe, 31, 67, 71–73, 76, 110, 119, 128–32, 144–48, 168, 179, 182, 195–98, 201, 203, 206, 207, 221, 222, 226
Ōji Paper Museum, 28–30, 35–38, 45–52, 60, 79, 80, 112, 117, 118, 125, 126, 140–42
Francis Haar, 61–63, 77, 78, 115, 116, 127, 143, 149, 167, 169, 170, 199, 200, 202, 204, 216, 225, 227
Mamoru Iwasaki, 89, 96–100, 111, 134–36
Naoki Baba, 64–66, 84–88, 178, 217–20
David Hughes, 39–44, 151–58
International Society for Educational Information, 1, 4, 5, 7–11, 13–15, 27
Masumi Hamada, 3